About the Author

Colin Beveridge is a maths confidence coach for Flying Colours Maths, author of *Basic Maths For Dummies* and *Numeracy Tests For Dummies* and co-author of the *Little Algebra Book*.

He holds a PhD in Mathematics from the University of St Andrews and worked for several years on NASA's Living With A Star project at Montana State University, where he came up with an equation which is named after him. It's used to help save the world from being destroyed by solar flares. So far so good.

He became tired of the glamour of academia and returned to the UK to concentrate on helping students come to terms with maths and show that not all mathematicians are boring nerds; some are exciting, relatively well-adjusted nerds.

Colin lives in Poole, Dorset with an espresso pot, several guitars and nothing to prove. Feel free to visit his website at www.flyingcoloursmaths.co.uk or follow him on Twitter at www.twitter.com/icecolbeveridge.

Dedication

In memory of my grandmother, Pat Knight, who taught me to be proud of being good with numbers.

Author's Acknowledgments

I'm grateful, as always, to the team at Dummies Tower for knocking my book into shape - particularly my editors Simon Bell, Kate O'Leary and Mike Baker, and my technical editors Giles Webberley and Alix Godfrey.

Thanks again to the Little Red Roaster in Parkstone for keeping me supplied with coffee and encouragement, and the students who helped me crystallise the ideas, especially Rebecca Murray and Jasmine Cooper.

And, of course, I thank my parents, who brought me up to say 'thank you' – Ken Beveridge and Linda Hendren.

Publisher's Acknowledgments

We're proud of this book; please send us your comments at http://dummies.custhelp.com. For other comments, please contact our Customer Care Department within the U.S. at 877-762-2974, outside the U.S. at 317-572-3993, or fax 317-572-4002.

Some of the people who helped bring this book to market include the following:

Acquisitions, Editorial, and Vertical Websites

Project Editor: Simon Bell

Commissioning Editor: Mike Baker

Assistant Editor: Ben Kemble

Development Editor: Kate O'Leary

Copy Editor: Kate O'Leary

Technical Editors: Giles Webberley, Alix Godfrey

Proofreader: Mary White

Production Manager: Daniel Mersey

Publisher: David Palmer

Cover Photos: © Shutterstock/ Robert Spriggs

Cartoons: Ed McLachlan

Composition Services

Project Coordinator: Kristie Rees

Layout and Graphics: Carl Byers, Carrie A. Cesavice, Corrie Niehaus, Christin Swinford

Proofreader: Lindsay Amones

Indexer: Potomac Indexing, LLC

Publishing and Editorial for Consumer Dummies

    Kathleen Nebenhaus, Vice President and Executive Publisher

    Kristin Ferguson-Wagstaffe, Product Development Director

    Ensley Eikenburg, Associate Publisher, Travel

    Kelly Regan, Editorial Director, Travel

Publishing for Technology Dummies

    Andy Cummings, Vice President and Publisher

Composition Services

    Debbie Stailey, Director of Composition Services

Contents at a Glance

Table of Contents

Chapter 14: Sharpening Your Knowledge of Shapes 241

Part V: The Part of Tens 325

Chapter 19: Ten (Or So) Ways to Check Your Work 327

Chapter 20: Ten Tips for Remembering
Your Number Facts 333

Introduction

*H*i! I'm Colin, and I want you to rock at maths.

I want you to have the skills you need to get the job you're going for or to qualify for a course or to help your kids with their homework. I want you to be able to say things like 'I use maths all the time . . . it's not that hard' or 'I'm pretty good at fractions'. And I want you to have the confidence to look at a sum the same way you look at a Sudoku or crossword puzzle: anticipating the pleasure of solving a problem.

You don't have to wear inch-thick glasses and a tweed jacket to do well enough at maths – and you don't need to have done well at school, either. Like anything else you want to master, you just have to approach the task with a positive attitude, a fair amount of stubbornness and detailed instructions. The first two are down to you; the third you'll find in this book!

I'm always keen to hear how you get on. The best way to reach me is via Twitter (@icecolbeveridge). I promise I'll read your comments and get back to you if I possibly can.

About This Book

This book is designed to help you get on top of the maths you need for the national numeracy curriculum at Entry Level 3, Level 1 and Level 2. It may help you prepare for a GCSE at foundation level as well, although several GCSE topics (notably algebra) aren't covered in this book at all.

This book is a companion to *Basic Maths For Dummies* (also written by me), which explains things in a bit more detail but doesn't contain practice questions.

I show you how to do the kinds of questions you're likely to come across in real life and in numeracy tests, particularly:

✔ Solving regular arithmetic problems – multiplying, dividing, adding and taking away.

✔ Rounding and estimating to get rough answers (or to check that your accurate answers make sense).

✔ Dealing with decimals, fractions and ratios.

✔ Messing about with measures of time, weight, temperature and money.

✔ Understanding shapes – measuring them, drawing them and moving them around.

✔ Grappling with graphs – how to read them and spot when someone's pulling the wool over your eyes.

✔ Summing up statistics, including averages and probabilities.

You have a lot to cover! But much of it is common sense, and if you can get past the – mistaken – idea that maths is really hard, you might even start to like it.

Conventions Used in This Book

I keep the conventions used in this book to a minimum! Here are the ones I use:

✔ I use *italics* for emphasis or to highlight new words and phrases.

✔ I use **boldface** for key words in bulleted lists or key steps in action lists.

✔ I use monotype font for Internet and email addresses.

What You're Not to Read

This book is designed to be an easy-access reference guide to basic maths. I cover each subject in its entirety in individual chapters, and the information doesn't depend on what comes before or after. This means you can jump around the book to the subjects you want to focus on and skip those you feel comfortable with already or just aren't interested in.

If you feel like you're starting from scratch, I strongly recommend you peruse the whole book to get an overview of all the subjects I cover. If you already have a decent maths background, you probably want to focus on the areas you still find challenging – but you may also find some of the insights in other areas help to shore up your maths skills.

No matter what your background, you can skip paragraphs marked with the Technical Stuff icon without foregoing an understanding of the primary subject. Sidebars likewise supplement the primary text – so you can skip them without missing the main point.

Foolish Assumptions

Making assumptions is always a risky business, but knowing where I'm coming from may put you at ease. So, in writing this book, I assume that:

- ✔ You know how to count and are familiar with the symbols for the numbers.
- ✔ You understand the idea of money and changing a banknote for an equivalent value of coins.
- ✔ You know what some basic shapes look like.
- ✔ You're prepared to think fairly hard about maths and want either to pass a numeracy examination or to simply brush up on your maths skills.

How This Book Is Organised

Like all *For Dummies* books, *Basic Maths Practice Problems For Dummies* is a reference book and each topic is allotted its own part in it. Within each part are individual chapters relating specifically to the topic in question.

Part 1: The Building Blocks of Maths

The first part of this book gives you the tools you need to get started with maths – if you can do basic arithmetic with whole numbers, it makes the rest of the book a lot easier! In this part, I show you how to:

- ✔ Add and take away whole numbers
- ✔ Multiply and divide whole numbers
- ✔ Estimate and round off approximate answers

Part 11: Working with Parts of the Whole

Stand back, everybody – I'm going to use the F-word. A word some people would like to see banned from books in public libraries and never have to hear on the TV. That's right, I'm talking about fractions!

I'm here to tell you that there's nothing dirty about fractions, even improper ones. Lots of misinformation exists about fractions and even some teachers find talking about them difficult.

But don't worry – in Part II I try to answer all of your questions about fractions in a frank and easy-to-understand manner. I cover their close friends, too – decimals, percentages, ratios and proportions, all versions of the same thing.

I also introduce you to the Table of Joy – an easy way to work with percentages, ratios and literally dozens of other topics. I use this table throughout the book. In fact, the Table of Joy is probably the most useful thing I know.

Part 111: Real-life Maths

The third part of this book is about applying your maths knowledge to real-life subjects – generally things that you measure.

Some of these concepts are perfectly familiar – you've probably worked with time and money since you were old enough to choose to be the top hat in a game of Monopoly. However, in some areas you need to be careful – and this book gives you a few extra tips and tricks for dealing with them.

Some of the measuring concepts are a bit trickier. I look at the different ways to measure weight and temperature and show you some of their many applications.

I also look at size and shape – again, different ways exist to measure these and you encounter many facets of shape to play with.

Part IV: Speaking Statistically

Statistics has a reputation for being boring and difficult. For a long while, I bought into that story, too – but then I started using statistics and applying the concept to something I cared about. Suddenly, I was drawing graphs that helped me understand my project, working out statistics that told me what was going on and making predictions based on probabilities . . . and I was hooked.

I can't promise you'll find statistics as exciting as I do, but I do my best to make the topic interesting. I cover the ins and outs and ups and downs of graphs and tables, how to interpret them and how to draw them; I look at averages; and I dip a toe into the murky and controversial world of probability.

Part V: The Part of Tens

All *For Dummies* books finish with these short, punchy chapters full of practical tips to help you manage the material in the rest of the book.

In this section, I show you how to check your work, remember your facts and make sure you get the right answer!

Icons Used in This Book

Here are the icons I use to draw your attention to particularly noteworthy paragraphs:

 Theories are fine, but anything marked with a Tip icon in this book tells you something practical to help you get to the right answer. These are the tricks of the mathematical trade.

 Paragraphs marked with the Remember icon contain the key takeaways from the book and the essence of each subject.

 The Warning icon highlights errors and mistakes that can cost you marks or your sanity – or both.

 You can skip anything marked with the Technical Stuff icon without missing out on the main message, but you may find the information useful for a deeper understanding of the subject.

Where to Go from Here

This book is organised so that you can jump right into the topics that interest you. If you feel like an absolute beginner in maths, I recommend you read Parts I and II to build a foundation for the other topics. If you're pretty comfortable with the mechanics of maths, use the Table of Contents and Index to find the subject you have questions about right now.

This book is a reference – keep it with your maths kit and turn to it whenever you have a question about maths.

Good luck!

Part I
The Building Blocks of Maths

"After reading that book on Basic Maths, I calculated we had more money that we thought."

In this part . . .

*I*n this part, I take you through the real foundations of Basic Maths: the skills you'll need to have so you can master the other, more involved chapters.

You're going to need to know how to add, take away, multiply and divide – as well as how to estimate and round numbers.

Once you've got all of that under your belt, the rest is easy!

Chapter 1

Getting Started

*Y*ou can do this.

Before you start, sit up straight, breathe in and take a minute to reassure yourself that you're smart and that you do maths all the time without realising it.

Every time you cycle to work, you perform feats of mathematics that would require supercomputers to work out in anything like the timescale your brain can do them in – from deciding which path to take to avoid the lorry, to figuring out exactly when to brake for the traffic lights, and even to remembering the combination for your bike lock.

Okay, maybe you don't need a supercomputer for the last one, but the point stands: you're much better at maths than you realise. Maybe you don't yet have a handle on the kind of maths you need to do well in exams, but that's just a matter of time.

In this chapter, I show you how to get better at that other kind of maths, the sort you need to get qualifications, and I take you quickly through the topics I cover in the rest of the book.

Covering the Basics

You may have a mental image of a mathematician – enormous forehead, crazy hair, thick glasses, tweed jacket over a tasteless shirt with pens neatly arranged in the breast pocket, gesticulating madly at a blackboard covered in crazy equations.

Actually, I do know mathematicians like that – but we're not all so poorly adjusted. Being good at maths doesn't automatically turn you into a socially awkward egghead.

That's not the only good news, though: you're also excused from having to understand all those crazy equations. Virtually no algebra is covered in the numeracy curriculum (just a few simple formulas). All you need to be able to do is:

- ✔ **Add, take away, divide and multiply confidently:** If you can use all of these maths tools, you'll probably find the chapters in Part I relatively easy to work through. If you can't, Chapters 2 to 4 help you build a solid foundation to work from.

- ✔ **Figure out the right sum to do:** Working out which tool to use to answer a question can be tricky, but if you keep a clear head and think through what the question is asking, it will make sense in the end. Promise.

- ✔ **Make sense of measures:** 'Measure' doesn't just mean being able to use a ruler, although that's a good starting point. It's also about weighing, taking temperatures, telling the time and working with shapes. For dealing with shapes, you just need to know a few simple formulas for area and volume.

- ✔ **Read and understand graphs and basic statistics:** Once you 'get' graphs, the answers start to jump off the page. You only need to care about a mere handful of types of graph, and you just need to figure out where each of them is hiding the information. Until you know that, graphs can be a bit confusing – but don't worry, I take you through them as gently as I can!

Talking Yourself Up

The stories you tell yourself are extraordinarily powerful. I used to tell myself I was useless and stupid, despite some evidence to the contrary. I was miserable, prone to panic attacks and generally conformed to what I'd told myself.

Eventually I made some changes to my life and told myself I was capable and intelligent instead. It was astonishing how quickly things got better – I still have the odd bad day, but at least I'm a functioning human being these days.

Unfortunately, the stories most people tell themselves about maths are just as poisonous as the stories I used to tell myself. So, before you get started, please do one thing for me: look at the stories you tell yourself. If you say things like 'I don't have a maths brain' or 'I'm rubbish at maths', you're digging a hole for yourself. Try telling yourself these stories instead:

> ✔ 'I used to struggle with maths – but I'm putting that right now.'
>
> ✔ 'I'm much better at maths than I thought!'
>
> ✔ 'I'm working on my maths skills.'

You don't need to tell yourself that you're going to win the next series of *Countdown* (although practising with games is a great way of honing your mental arithmetic), just give yourself a good name to live up to!

Collecting the tools you need

A very popular joke among mathematicians states that maths is the second-cheapest subject to study because all you need is a pencil, some paper and a bin. Philosophy, of course, is cheaper because you don't need the bin.

For this book, you can do an awful lot with just the pencil, paper and bin, but you may also find a few other bits and pieces useful, too:

> ✔ **A calculator:** While you don't *need* a calculator for working through this book, it's quite a useful thing to have around. Most numeracy tests are non-calculator papers, but you're allowed to study however you like. If you want to use a calculator, go ahead – just don't rely on it!

✔ **A dedicated notebook or folder:** This advice falls into the 'do as I say, not as I do' category – my notes are scattered all over my flat. The upshot is that I can never find anything I'm working on – and I don't want that to happen to you! Keeping your notes in one place makes reviewing them a lot easier.

✔ **A geometry set:** A ruler and a protractor may prove very useful – and if you're going to have those, why not a compass, a set square and the stencil nobody ever uses?

✔ **A comfortable, quiet place in which to work:** Working through maths problems is harder if you're distracted. If at all possible, find a space in which you can sit comfortably without anyone bothering you for a while and work there.

Handling Whole Numbers

Whole numbers are the building blocks of maths. Pretty much anything you do in basic maths requires you to have a good handle on them. You need to be good at three sets of tools:

✔ **Adding and taking away:** Adding up is probably the first thing you learn in maths after counting; taking away is a little more difficult, but not by much. If you can count, you can add and take away – and I show you how in Chapter 2.

✔ **Multiplying and dividing:** These tools are slightly more difficult than adding and taking away; most people find multiplying ('timesing') a bit easier than dividing. I give you simple, reliable methods that make both of them straightforward in Chapter 3.

✔ **Rounding and estimating:** In some ways, these tools are the most important. They stop you being overcharged in the supermarket and getting run over as you cross the road. I show you how to get rough answers and how to round off in Chapter 4.

Dealing with Parts of Numbers Basic maths involves some work on things smaller than whole numbers. That means . . . yes, fractions and decimals. Oh, and percentages, too. Look, don't blame me, I'm just the messenger. And fortunately I have some good news: I introduce you to the Table of Joy, which makes ratios and percentages (and all manner of conversions) as easy as pie.

The chapters in Part II cover the following methods for dealing with part numbers:

- ✔ **Fractions:** In basic maths you only need to find fractions of a whole number, cancel fractions down (and up) and add and subtract fractions. Chapter 5 covers all of these concepts. Once you get the trick, dealing with fractions is easy – honestly.

- ✔ **Decimals:** You're probably a bit more familiar with decimals – after all, most prices contain decimal points. The rules of arithmetic are no different for decimals than for whole numbers; you just need to keep your eye on the dot! I run you through the methods for working with decimals in Chapter 6.

- ✔ **Ratios and percentages:** Dealing with ratios (Chapter 7) and percentages (Chapter 8) is a big favourite of maths examiners, presumably because they sometimes come up in real life. I show you how to use the Table of Joy to figure out which sum to do for both of these – soon you'll be doing them in your sleep!

Managing Measurements

By measurements, I don't just mean using a tape measure, although measuring distance is part of this subject. 'Measurements' could just as easily be called 'real-life maths' because it deals mainly with using your knowledge from Parts I and II to solve problems in the outside world. In more detail, I cover:

- ✔ **Time:** You probably have a decent idea of how time works but, all the same, I take you through the different ways of telling time and working with timetables and other time sums in Chapter 9.

- ✔ **Money:** Money is probably the bit of basic maths you use most in your everyday life. In Chapter 10, I show you how to deal with money sums – which work just like normal sums – and deal with more complicated things such as deposit schemes and exchanging currencies.

✔ **Weight:** There's not all that much to say about weight in Chapter 11, except that the sums you do work just like any other kind of sum. The only tricky bit might be converting between different units, but you have the Table of Joy for that!

✔ **Temperature:** Temperature, which you read about in Chapter 12, is where things can get a bit tricky. You have to deal with negative numbers and possibly some formulas for converting between different temperature scales. Here's some good news, though: temperature is almost always in whole numbers!

✔ **Size and shape:** In Chapters 13 and 14 you find out about the difference between length, area, volume and angle, and how to figure each of them out!

Speaking statistically

In Part IV, I give you a very brief introduction to *statistics*, which is just using numbers to summarise data. Basic maths barely scratches the surface of statistics, which I used to hate; however, when I started using it for something actually relevant to me, it made much more sense! In this part, I tell you about:

✔ **Graphs and tables:** In Chapters 15 and 16, you see how to read values from graphs and tables, and do more complicated sums with the results. I also give you a few ideas about drawing your own graphs.

✔ **Averages and spread:** Chapter 17 is the place to go if you want to be able to tell your mean from your median and your mode! Working out the different kinds of average is a very typical exam question. Don't worry, though; when you've completed the problems in this chapter, you'll find working out averages and spread a walk in the park.

✔ **Probability:** In Chapter 18, I show you the basics of probability. As the word suggests, this maths tool is all about working out how likely something is to happen.

Working through Questions and Answers

The whole point of this book is to help you get better at basic maths by following examples and answering sample questions.

Basic Maths Practice Problems For Dummies is organised so that each part builds on the parts before. If you feel that you need to start from scratch, starting at the beginning and working your way forward probably makes sense.

You can study however you like, though! Don't feel that you have to run through the book from beginning to end; you're perfectly free to jump around from chapter to chapter, or simply to do a handful of questions picked at random from anywhere in the book.

Chapter 2

Introducing the Basics: Addition and Subtraction

*F*irst, you learn to count. Then you discover adding and taking away (*subtracting*). These sums are the ones you do most often in your day-to-day life. Just off the top of my head, I expect to use adding up today to:

- ✔ Work out my shopping bill.

- ✔ Tot up how many hours I've worked today.

- ✔ Get to the bottom of my killer Sudoku.

I expect to use taking away to:

- ✔ Work out what time I need to leave to get to my class on time.

- ✔ Check how much change I ought to get when I buy my groceries.

- ✔ See how many points away from winning the league my team is (clue: it's an awful lot; we're bottom at the moment).

In this chapter, I show you how to add and subtract with confidence. I start with the basic facts you need to know and then go on to the simple methods for adding and taking away. After that, I cover what to do when the simple methods are just a bit too simple, and then spend a bit of time on the number line and negative numbers so that they're no longer a mystery to you.

Adding Up Small and Large Numbers

You *add* things when you have two amounts you want to combine together. For instance, if you have a tin of beans (35p) and a loaf of bread (120p) and want to find out how much they cost altogether, you add up the pair of numbers. (The answer is 155p, but I haven't shown you how to do that yet.)

Words such as 'altogether', 'total' and 'sum' provide you with a clue that you may have to add things up.

This section starts with small numbers and gradually works up to adding up big numbers. I start with the basic facts (adding up numbers from 1 to 10), then show you how to add numbers bigger than 10 and finally show you what to do when your numbers get *too* big. (The answer is, you have to *carry* the too-big number, which is a much less complicated process than it sounds.)

Starting with 1 to 10

To be able to add up confidently, you need to know how to add up the numbers from 1 to 10 almost without thinking about them. Figure 2-1 is a classic adding-up grid, which allows you to look up the numbers while you're learning.

Follow these steps to use the grid to add two numbers together:

1. **Find the first number in your sum in the top, horizontal, row.**

2. **Find the second number in your sum in the left-hand, vertical, row.**

3. **Run one finger down the grid from the top, horizontal, number and, with a different finger, across from the number you chose in the vertical column until your fingers meet.**

   The number in this square is your answer.

+	0	1	2	3	4	5	6	7	8	9	10
0	0	1	2	3	4	5	6	7	8	9	10
1	1	2	3	4	5	6	7	8	9	10	11
2	2	3	4	5	6	7	8	9	10	11	12
3	3	4	5	6	7	8	9	10	11	12	13
4	4	5	6	7	8	9	10	11	12	13	14
5	5	6	7	8	9	10	11	12	13	14	15
6	6	7	8	9	10	11	12	13	14	15	16
7	7	8	9	10	11	12	13	14	15	16	17
8	8	9	10	11	12	13	14	15	16	17	18
9	9	10	11	12	13	14	15	16	17	18	19
10	10	11	12	13	14	15	16	17	18	19	20

Figure 2-1: The adding-up grid.

Knowing these simple additions by heart really makes working out harder sums an awful lot easier! You can practise them with a pack of cards, as follows:

1. **Deal two cards face up, side by side.**

   If you turn over a jack, queen or king, deal another card on top of it.

2. **Add the numbers on the two cards together in your head as quickly as you can.**

3. **Check your answer in the adding-up grid.**

   Repeat Step 1 until you've worked through all the cards.

To spice things up, you can go through the pack against the clock and give yourself a time penalty for every addition you get wrong. Try to beat your best time every time you play.

Working up to bigger numbers

Adding bigger numbers isn't that much harder. In this recipe, I use the word *number* to mean the numbers you're adding together (for instance, in 235 + 594, 235 and 594 would be the numbers) and the word *digit* to describe the symbols that make up the numbers (so the *number* 235 is made up of the *digits* 2, 3 and 5).

Here's a recipe for how you work out 235 + 594:

1. **Write the numbers down, one above the other, so that the last digits are in line.** It's okay if you have a space at the start of one of the numbers, but not in the middle or at the end. Draw a line under the last number. (See Figure 2-2(a).)
2. **Start with the right-hand column and add up all the digits in it.** (In this case, 5 + 4 = 9.)
3. **If the answer is less than 10, write it beneath the underline, in the same column that you just added up.** (See Figure 2-2(b).)
4. **If the answer from Step 2 is more than 10, write the second digit of the answer beneath the underline in the same column.** Write the first digit below the next column to the left – doing so is called *carrying*. (See Figure 2-2(c).) You need to add the carried digit to the numbers in the column which it is now beneath.
5. **Repeat Steps 2 to 4 for each column in turn, working from right to left.** The number beneath the underline is the answer. (Figure 2-2(d) provides the final answer, which is 829.)

To recap the method shown in Figure 2-2: I added the digits 5 and 4 from the right-hand column to get the 9 at the right-hand end of my answer. Then I added the digits 3 and 9 from the middle column to get 12 – I wrote 2 underneath the column and 1 below the left-most column. Then I added the digits 1, 2 and 5 to get 8, my answer for the left-most column. My answer is 829.

235	235	235	235
+ 594	+ 594	+ 594	+ 594
	9	29	829
		1	1
a.	b.	c.	d.

Figure 2-2: Adding up 235 + 594.

Attempting some adding questions

Test your knowledge of addition by working through these quick questions.

1. Work out:

(a) 5 + 3   (b) 1 + 9   (c) 3 + 1   (d) 4 + 7

(e) 4 + 10   (f) 6 + 8   (g) 7 + 9   (h) 9 + 8

2. Work out:

(a) 88 + 11   (b) 34 + 55   (c) 23 + 76   (d) 11 + 20

(e) 60 + 57   (f) 35 + 76   (g) 97 + 9   (h) 61 + 47

3. Work out:

(a) 167 + 717   (b) 624 + 316   (c) 417 + 551   (d) 575 + 144

(e) 462+146   (f) 711 + 661   (g) 533 + 494   (h) 956 + 301

4. A decathlete has 6504 points as she lines up for the final event (the 1500 metres). She scores 758 points. What is her total score?

The answers are provided at the end of the chapter.

Subtracting Small and Large Numbers

You *subtract* things, or *take them away*, when you want to find the difference between two amounts, or how much is left after you use a certain amount. For instance, if you buy a 55p chocolate bar using a pound coin (100p), you do a take-away sum to find out how much change you should get. (The answer is 45p.)

Words such as 'difference' or 'change' provide a clue that you may have to take things away.

This section first shows you how to subtract small numbers and then, when you've mastered the basics, leads you through taking away big numbers. I show you two ways to take away: the traditional 'column' method you probably used at school and got confused by, and a much better 'do-the-same-thing' method that makes the sums much easier!

Starting with 1 to 20

Figure 2-3 is a classic grid for working out subtractions using the column method. (One of the many things I prefer about the do-the-same-thing method is that you don't need to learn how to use a grid or to work with columns!)

	0	1	2	3	4	5	6	7	8	9	10	11	12	13	14	15	16	17	18	19	20
0	0	1	2	3	4	5	6	7	8	9	10	11	12	13	14	15	16	17	18	19	20
1		0	1	2	3	4	5	6	7	8	9	10	11	12	13	14	15	16	17	18	19
2			0	1	2	3	4	5	6	7	8	9	10	11	12	13	14	15	16	17	18
3				0	1	2	3	4	5	6	7	8	9	10	11	12	13	14	15	16	17
4					0	1	2	3	4	5	6	7	8	9	10	11	12	13	14	15	16
5						0	1	2	3	4	5	6	7	8	9	10	11	12	13	14	15
6							0	1	2	3	4	5	6	7	8	9	10	11	12	13	14
7								0	1	2	3	4	5	6	7	8	9	10	11	12	13
8									0	1	2	3	4	5	6	7	8	9	10	11	12
9										0	1	2	3	4	5	6	7	8	9	10	11
10											0	1	2	3	4	5	6	7	8	9	10

Figure 2-3: The taking-away grid.

Follow these steps to use the grid in Figure 2-3 to work out 12 – 7:

1. **Find the first number (12) in the top row.**

2. **Find the second number (7) in the left-hand column.**

3. **Run a finger down from the top row and a finger along from the left-hand column until they meet.**

The number in that square is your answer.

But what if your second number is bigger than your first number and you find a blank space where your answer ought to be? Well, it depends on what you're doing:

- ✔ If the first number of the whole sum you're working out is smaller than the second, you need negative numbers. Take a look at the 'Getting to grips with negative numbers' section later in this chapter.

- ✔ If you're working out a sum using the column method, you need to borrow. This technique is explained in the 'Using the column method' section later in this chapter.

Working up to bigger numbers

In this section, I show you two different methods for taking away bigger numbers. If you have a vague recollection of taking things away using the column method at school (if Figure 2-4 below rings bells with you and you just need to refresh a few things), you'll probably prefer this approach. If you just remember taking away being hard and complicated, you may prefer to skip over the column method and just learn the 'do-the-same-thing' approach.

The choice of method is always up to you. As long as it consistently gets you the right answer, I don't mind how you get there.

Using the column method

This is almost certainly the method you were taught at school. In this recipe, I show you how to work out 849 – 564. You can follow the stages in Figure 2-4(a–e). Follow these steps:

			⁷¹	⁷¹
849	849	8̶49	8̶49	8̶49
– 564	– 564	– 564	– 564	– 564
a.	b. 5	c. 5	d. 85	e. 285

Figure 2-4: Working out 849 – 564 using the column method.

1. **Write down the first number above the second number, and line them up so that the final digits are in the same column.**

Give yourself plenty of space around the digits and try to make it obvious which column is which.

2. **Start from the right-hand column.**

3. **If the top digit is bigger than the bottom digit, take the bottom digit away from the top one and write it at the bottom of the column.**

   Here, in the right-hand column, you write 9 − 4 = 5.

4. **If the bottom number is bigger than the top one, you need to *borrow* from the next column to the left.** That means, take 1 away from the top number to the left, cross that number out and write the new number in its place; then add 10 to the top number you're working with. In the second column, you turn the 8 in the column on the left into a 7, and the 4 you're working on becomes 14. Then take the bottom number from the new top one and write it below: 14 − 6 = 8. Phew!

5. **Repeat Steps 3 and 4, working from right to left.** The number underneath at the end is the answer: 285.

Borrowing excessively

You may ask a really good question here: what happens if there's nothing in the next column to borrow from? For instance, if you want to work out 100 − 9, you look at the right-hand column and think, 'I can't take 9 away from 0, so I need to borrow from the next column to the left . . . but that's zero as well!'

You're quite right, you can't borrow from a number that doesn't have anything in it. The solution is a little bit complicated, but here's what you need to do. If the number you want to borrow from is zero, you need to fix that. You can turn it into 10, though, if you borrow 1 from the column to

its left. Now you have something to borrow from.

In the example, 100 − 9, you can't take 9 away from 0, so you try to borrow from the middle 0 − but obviously you can't. You need to turn the middle column into 10 by borrowing one from the left-most column (which becomes 0). Now you can borrow from the middle column (which becomes 9) and the right-hand column becomes 10.

This borrowing means you can do the take-away sum easily now: the right-hand column is 10 − 9 = 1; the middle column is 9 − 0 = 9 and the left-hand column is empty. The answer is 91. The figure below takes you through the stages of working out both this problem and one other.

$$
\begin{array}{cccc}
 & {}^{0\,1} & {}^{0+9\,1} & {}^{0+9\,1} \\
1\,0\,0 & +0\,0 & +0\,0 & +0\,0 \\
\underline{\quad 9} & \underline{\quad 9} & \underline{\quad 9} & \underline{\quad 9} \\
 & & & 0\,9\,1
\end{array}
$$

$$
\begin{array}{ccccc}
 & {}^{0\,1} & {}^{0+9\,1} & {}^{0+9+9\,1} & {}^{0+9+9\,1} \\
1\,0\,0\,0 & +0\,0\,0 & +0\,0\,0 & +0\,0\,0 & +0\,0\,0 \\
\underline{\quad 5\,4} & \underline{\quad 5\,4} & \underline{\quad 5\,4} & \underline{\quad 5\,4} & \underline{\quad 5\,4} \\
 & & & & 0\,9\,4\,6
\end{array}
$$

Doing the same thing

You may be thinking, 'There has to be an easier way'. And you're absolutely right. The do-the-same-thing method involves (as you may suppose) doing the same thing to both of the numbers. With this approach, you can do a take-away sum with hardly any actual taking away!

Using the do-the-same-thing method means that whatever you do to one number (adding something on or taking something away), you have to do the same to the other number. If you're playing a game and you and your opponent both score, say, 20 more points, the difference between your scores is the same as before. You can use this to make the second number, the one you take away, as easy as possible to deal with.

For example, say you need to work out 1000 – 54. Here's how you do it:

1. **Find the last digit of the second number that isn't zero.**

2. **Work out what you need to add on to turn that digit into a zero, and add it on.** (In the example, to turn 54 into 60, you add 6; to turn 60 into 100, you add 40; and to turn 100 into 1000, you add 900.)

3. **Add the same amount to the first number.**

4. **If the take-away sum is easy, do it (using the column method if you need to).** Otherwise, go back to Step 1 until the sum becomes easy.

So, to work out 1000 – 54, I say to myself:

> ✔ The last digit of 54 that isn't zero is 4, so I'll add 6 to both numbers and the sum becomes 1006 – 60.
>
> ✔ The last digit of 60 that isn't zero is 6, so I'll add 40 to both numbers and the sum becomes 1046 – 100.
>
> ✔ The last digit of 100 that isn't zero is 1, so I'll add 900 to both numbers and the sum becomes 1946 – 1000.
>
> ✔ That sum is easy! The answer is 946.

Attempting some subtracting questions

1. Let's start with some simple subtractions:

(a) 9 – 5   (b) 8 – 3   (c) 7 – 2   (d) 9 – 2

(e) 10 – 8   (f) 11 – 2   (g) 15 – 8   (h) 13 – 9

(i) 12 – 6   (j) 11 – 9   (k) 12 – 7   (l) 10 – 6

2. Now, using either the column or the do-the-same-thing method, work out some two-digit subtractions:

(a) 99 – 43   (b) 69 – 13   (c) 95 – 30   (d) 46 – 15

(e) 92 – 85   (f) 82 – 28   (g) 45 – 16   (h) 80 – 66

3. And now for some even bigger numbers:

(a) 1340 – 966   (b) 746 – 205   (c) 1009 – 60

(d) 1248 – 860   (e) 931 – 611   (f) 826 – 285

4. In Week 1, 9713 drivers use a particular car park. In Week 2, 208 fewer drivers park there. How many drivers used the car park in Week 2?

The answers are provided at the end of the chapter.

Doing Sums with a Number Line

The *number line* is a marvellous tool for working out adding and taking away problems. It is just a line with numbers on it that looks a bit like a ruler – see Figure 2-5. Notice how it goes below zero? Ignore that for the moment because I tell you all about it in the 'Getting to grips with negative numbers' section below.

−15 −14 −13 −12 −11 −10 −9 −8 −7 −6 −5 −4 −3 −2 −1 0 1 2 3 4 5 6 7 8 9 10 11 12 13 14 15

Figure 2-5: The number line from −15 to 15.

I don't recommend adding and taking away with the number line as an all-purpose tool, because the other methods I describe in this chapter are almost always easier to work with, but it is *really* useful for dealing with negative numbers. So, to start with, here's how you add and take away normal (*positive*) numbers with the number line:

1. **Put your finger on the first number in the sum.**

2. **Identify the second number in the sum.** If you're doing an adding sum, move your finger that many spaces to the *right*. If you're doing a take away sum, move your finger that many spaces to the *left*. (I remember to go left for a take away sum by saying to myself 'less is left'.)

3. **The number you land on is your answer.**

Now check your understanding by using Figure 2-5 and a few facts you already know: to work out 3 + 6, you put your finger on 3, move it six spaces to the right and you end up on 9, which is the right answer. To work out 12 − 5, you put your finger on 12 and move it five spaces to the left. You end up on 7, which is also the right answer.

Getting to grips with negative numbers

Normal numbers – ones that are bigger than zero – are called *positive* numbers. But you also get *negative* numbers, which are less than zero and are written with a minus sign in front of them, for example –7. So, when the temperature is colder than freezing, the weather report gives you a *negative* temperature. If your football team has let in more goals than it's scored (like my team almost always does), it has a *negative* goal difference.

Negative numbers are also used for showing changes and differences. If you look at a football league table, especially near the bottom, you'll see that some teams have a negative goal difference – they've let in more goals than they've scored. In other tables, you might show a change in a value with a + if it's gone up and and – if it's gone down. For example, if a train company's fares had dropped by an average of 5 per cent (chance would be a fine thing!), a table might show its fare increase as –5 per cent. Dealing with money is the same – you can show a *loss* instead of a profit by putting a minus sign in front of the number. If I lose £25 in a casino, I can claim that I actually won –£25! I'm not sure anyone would be all that impressed, though.

Only a few kinds of negative number sums actually come up in basic maths tests; these are:

✔ Finding the difference between two negative numbers

✔ Finding the difference between a negative number and a positive number

✔ Adding numbers on to a negative number

✔ Taking numbers away from a negative number

✔ Putting positive and negative numbers into size order

The following sections show you how you do each of them.

Finding the difference

When you find the difference between two normal, positive numbers, you can think about how much you have to change the first one to get to the second. If you're working out 12 – 7, you may say, 'I need to count down by five from 12 to get to seven, so the difference is five.'

You can do a very similar thing with negative numbers. Finding the difference between two negative numbers is really easy: you simply ignore the minus sign altogether and take away as normal. The difference between –7 and –12 is 5, just like in the example before. If you were thinking about temperatures, you'd say that $-7°$ was five degrees warmer than $-12°$ (although it's still pretty nippy if you ask me!).

The situation's a bit more tricky if one of the numbers is positive and the other negative, but not much: the trick is to ignore the minus sign, and then *add* to get the answer. So the difference between $-12°$ and $7°$ is $12 + 7 = 19°$. And that makes sense: to get from -12 to 0, you'd have to count up 12, and to get from 0 to 7, you'd have to go up another 7, making 19 altogether.

Adding and taking away from a negative number

One way to add and take away from negative numbers is to use the number line. Here's how it works:

1. **Draw a number line like the one shown in Figure 2-5.**

2. **Put your pencil on the number you're starting from (it'll be to the left of zero).**

3. **If you're adding a number on, move your pencil that many spaces to the *right*. The number you land on is your answer.**

4. **If you're taking a number away, move your pencil that many spaces to the *left*. The number you land on is your answer.**

This approach is fine when you have small numbers, but you don't want to be drawing out a number line to deal with huge numbers – you'd be there all day! Fortunately, a method exists so that you can add and subtract negative numbers without the help of the number line.

The trick for adding on to a negative number is to *turn the sum around*. Here's what I mean by that: $-2 + 5$ is the same as $5 - 2$. Similarly, $-10 + 6$ is the same as $6 - 10$. So here's a recipe for doing just that (and working out the sum – in this case, let's do $-7 + 3$):

1. **Write down the number you have to add on (the one after the +). Here, you write down 3.**

2. **Write down the number you started with (including the minus sign) to make a take away sum. Now you have $3 - 7$.**

3. **If the first number in the take away sum is bigger than the second, great!** You can do it the normal way and get your answer.

4. **If the second number is bigger, swap the numbers around (here, you get to 7 – 3).**

5. **Work out this sum and put a minus sign in front of it: –7 + 3 = –4.**

Taking away from a negative number is even easier! It effectively turns into an add sum. If you need to do –15 – 6, you'd follow these steps:

1. **Ignore the minus signs and write down the numbers with a plus between them. Here, you get 15 + 6.**

2. **Work out the sum. Here, you have 21.**

3. **Put a minus sign in front. The answer is –21.**

Thinking about whether the sum would make the temperature warmer or colder is a good way to check whether your answer has gone the right way!

Putting numbers in order

You may be asked to put a list of numbers – including negative numbers – into the correct order. Although this task may seem complicated, it's actually really easy. Here are the steps:

1. **Split the list up into positive numbers and negative numbers.** Keeping them separate is vital!

2. **Sort the positive numbers into size order, smallest number first.**

3. **Sort the negative numbers into size order, putting the one that's closest to zero *last*.** (You can ignore the minus sign while you're sorting these.)

4. **Write down the sorted negative list from Step 3, followed by the sorted positive list from Step 2.** They're now in the right order.

If you've organised the numbers correctly, the list should start by moving towards zero (ignoring the minus signs, the numbers look like they're getting smaller) and then away from zero as you get into the positive numbers.

As an example, if you had to sort the numbers –4, 29, –26, 10, –17 and –12 into order, you'd split off the positive numbers (29 and 10) from the negative numbers (–4, –26, –17 and –12).

You'd sort the positive numbers in increasing order (10, 29) and the negative numbers (–26, –17, –12, –4) so that the number nearest zero was at the end. Then you'd join the lists together to get –26, –17, –12, –4, 10, 29. If you ignore the minus signs, the numbers get smaller until you get to zero, then they get bigger again.

Tackling some negative number questions

1. Find the difference between:

(a) –3 and –5 (b) –3 and –8   (c) –5 and –7   (d) –10 and –7

(e) –1 and –3 (f) –10 and –3   (g) –17 and –12 (h) –100 and –10

2. Find the difference between:

(a) –3 and 5   (b) –3 and 8   (c) 5 and –7 (d) 10 and –7

(e) –1 and 3   (f) –10 and 3   (g) 17 and –12   (h) 100 and –10

3. Work out:

(a) –3 + 5   (b) –5 + 3   (c) –7 + 10   (d) –6 + 3

(e) –20 + 7   (f) –20 + 30   (g) –15 + 10   (h) –10 + 10

4. Work out:

(a) –7 – 5   (b) –3 – 10   (c) –12 – 4   (d) –10 – 4

(e) –19 – 1   (f) –100 – 5   (g) –3 – 5   (h) –30 –50

5. Put the following sets of numbers into order:

(a) –9, –14, 4, –16, 7

(b) –13, 13, 15, –9, 7

(c) 19, –7, –15, –16, 4, –6, 15, –17

(d) –43, –26, –19, 32, 33

The answers are provided at the end of the chapter.

Working through Review Questions

1. In your head, work out:

(a) 1 + 2   (b) 1 + 8   (c) 3 + 5   (d) 8 + 2

(e) 3 + 6   (f) 4 + 4   (g) 4 + 7   (h) 8 + 5

(i) 8 + 8   (j) 9 + 7   (k) 9 + 5   (l) 7 + 8

2. Work out:

(a) 51 + 20   (b) 63 + 14   (c) 77 + 5   (d) 50 + 26

(e) 27 + 27   (f) 57 + 60   (g) 67 + 43   (h) 84 + 6

(i) 638 + 396   (j) 320 + 186   (k) 992 + 391   (l) 785 + 921

3. In your head, work out:

(a) 3 – 2   (b) 5 – 3   (c) 6 – 6   (d) 9 – 3

(e) 12 – 5   (f) 12 – 2   (g) 11 – 4   (h) 15 – 6

(i) 15 – 8   (j) 16 – 7   (k) 16 – 8   (l) 13 – 7

4. Work out:

(a) 48 – 5   (b) 59 – 27   (c) 85 – 13   (d) 37 – 29

(e) 72 – 5   (f) 84 – 77   (g) 83 – 28   (h) 70 – 38

(i) 357 – 72   (j) 979 – 359   (k) 748 – 168   (l) 314 – 85

5. Work out:

(a) 3 – 5   (b) –13 + 15   (c) –12 + 18   (d) –7 – 7

(e) –6 – 3   (f) 8 –11   (g) –18 – 13   (h) –3 + 9

(i) –16 + 11   (j) –1 + 17   (k) 10 – 18   (l) –5 – 1

The answers are provided at the end of the chapter.

Checking Your Answers

If you can get through all of these, you're well on the way to doing brilliantly at basic maths!

Adding up and taking away are two of the skills you use most often in almost any kind of maths, so it's worth practising these until you can get them right every time – almost automatically.

Making the occasional mistake is okay – I quite often mess up because I'm hurrying or not paying enough attention. The trick is to catch your mistakes and put them right!

Adding questions

1. (a) 8   (b) 10   (c) 4   (d) 11   (e) 14   (f) 14
   (g) 16   (h) 17

2. (a) 99   (b) 89   (c) 99   (d) 31   (e) 117   (f) 111
   (g) 106   (h) 108

3. (a) 884   (b) 940   (c) 968   (d) 719   (e) 608   (f) 1372
   (g) 1027   (h) 1257

4. 6504 + 758 = 7,262

Subtracting questions

1. (a) 4   (b) 5   (c) 5   (d) 7   (e) 2   (f) 9
   (g) 7   (h) 4   (i) 6   (j) 2   (k) 5   (l) 4

2. (a) 56   (b) 56   (c) 65   (d) 31   (e) 7   (f) 54
   (g) 29   (h) 14

3. (a) 374   (b) 541   (c) 949   (d) 388   (e) 320   (f) 541

4. 9713 – 208 = 9,505

Negative number questions

1. (a) 2   (b) 5   (c) 2   (d) 3   (e) 2   (f) 7
   (g) 5   (h) 90

2. (a) 8   (b) 11   (c) 12   (d) 17   (e) 4   (f) 13
   (g) 29   (h) 110

3. (a) 2   (b) –2   (c) 3   (d) –3   (e) –13   (f) 10
   (g) –5   (h) 0

4. (a) –12   (b) –13   (c) –16   (d) –14   (e) –20   (f) –105
   (g) –8   (h) –80

5. (a) –16, –14, –9, 4, 7   (b) –13, –9, 7, 13, 15   (c) –17, –16, –15, –7, –6, 4, 15, 19   (d) –43, –26, –19, 32, 33 (already in order!)

Review questions

1. (a) 3   (b) 9   (c) 8   (d) 10   (e) 9   (f) 8
   (g) 11   (h) 13   (i) 16   (j) 16   (k) 14   (l) 15

2. (a) 71   (b) 77   (c) 82   (d) 76   (e) 54   (f) 117
   (g) 110   (h) 90   (i) 1034   (j) 506   (k) 1383   (l) 1706

3. (a) 1   (b) 2   (c) 0   (d) 6   (e) 7   (f) 10
   (g) 7   (h) 9   (i) 7   (j) 9   (k) 8   (l) 6

4. (a) 43   (b) 32   (c) 72   (d) 8   (e) 67   (f) 7
   (g) 55   (h) 32   (i) 285   (j) 620   (k) 580   (l) 229

5. (a) –2   (b) 2   (c) 6   (d) –14   (e) –9   (f) –3
   (g) –33   (h) 6   (i) –5   (j) 16   (k) –8   (l) –6

Chapter 3

Equal Piles: Multiplying and Dividing

*M*ultiplying and dividing are a step up from adding and taking away. You multiply all the time without really thinking about it – if you know one bag of oranges costs £2, then three bags are $3 \times £2 = £6$.

You probably divide less often – but any time you split something up fairly, you're doing a divide sum.

In this chapter, I show you how to divide and multiply – starting with the simple facts and gradually working up to bigger numbers and the bogeyman that is long division. Division is quite simple as soon as you know what you're doing – promise!

Getting on the Grid: Multiplying the Easy Way

Multiplication is a short-cut to adding the same thing over and over again. If you have five bags, each of which contains six oranges, and you want to know how many oranges you have altogether, you could either work out:

$$6 + 6 + 6 + 6 + 6 = 30$$

or you could just say $5 \times 6 = 30$ straight away. When you only have five of something, adding repeatedly isn't a big task, but what if you have seven? Or twelve? Or a hundred? The bigger the number, the more work you save by being able to multiply.

In this section, I show you the number facts you need to know if you want to multiply well, take you through the way you can easily multiply by 10 (and 100 and so on) and demonstrate how you can apply those things to multiplying with a grid – which you can extend to multiplying numbers as big as you like.

Memorising your times tables

If you want to be able to multiply well, you need to memorise your times tables. Yes, I know, it's boring and time consuming and you have a calculator for that kind of thing, but knowing your times tables well will pay off a hundred times over when it comes to maths exams. Figure 3-1 shows the times tables in all their glory.

x	0	1	2	3	4	5	6	7	8	9	10
0	0	0	0	0	0	0	0	0	0	0	0
1	0	1	2	3	4	5	6	7	8	9	10
2	0	2	4	6	8	10	12	14	16	18	20
3	0	3	6	9	12	15	18	21	24	27	30
4	0	4	8	12	16	20	24	28	32	36	40
5	0	5	10	15	20	25	30	35	40	45	50
6	0	6	12	18	24	30	36	42	48	54	60
7	0	7	14	21	28	35	42	49	56	63	70
8	0	8	16	24	32	40	48	56	64	72	80
9	0	9	18	27	36	45	54	63	72	81	90
10	0	10	20	30	40	50	60	70	80	90	100

Figure 3-1: The times tables grid.

Here's how you use the times tables grid to multiply two numbers (up to 10 × 10) together:

1. **Find the first number in the top row of the grid.**

2. **Find the second number in the first column of the grid.**

3. **Read down the column in Step 1 and across the row in Step 2 until you find the number that's in both. This number is your answer.**

For instance, to work out 6 × 7, you read down the column with 6 at the top until you reach the row with 7 at the start and read off the answer: 42.

Timesing by ten

Multiplying by 10 is probably the easiest sum you'll ever be asked to work out. Maybe you've heard that you just stick a zero on the end – that's true for whole numbers, but it messes you up when you come to work with decimals, so here's a slightly different way to do it (but that works with any number):

1. **Write down the number you want to times by 10.**

2. **If it doesn't have a dot in it, put a dot at the end.**

3. **Move the number one space to the left, but leave the dot where it is.**

4. **Fill in any gaps with 0. This number is your answer.**

If you use the above method for multiplication, you won't have any trouble when it comes to Chapter 6 on decimals – but if you prefer to just stick a zero on the end, I'm not going to stop you!

You use the same method for multiplying by 100 or 1,000 but rather than moving one place to the left, you move two or three spaces instead – or however many zeroes are in the number.

You may also need to multiply by a number like 20 or 400 or 7,000 – a number followed by a string of zeroes. To work out a sum like that, for example $3 \times 7,000$, follow these steps:

1. **Multiply your number by the number at the start: $3 \times 7 = 21$.**

2. **Put a dot at the end of your answer.**

3. **Move your answer left by as many spaces as you have zeroes; for this sum, three.**

4. **Fill in the gaps with zeroes. Here, you get 21,000.**

Drawing the grid

Rather than the method I describe in the previous section, you may have learnt to multiply numbers in a different way at school. It does the same thing, though.

It doesn't matter what method you use to get the right answer. As long as it works for you, no one's going to yell at you for using a different method. Or, if they do, tell them I said you could do it your way as long as you get the right answer.

Here's how to work out any number times by a one-digit number. In the next section, you can discover how to multiply numbers of any length together – but you need to walk first before you can run. How about 234×7?

1. **Draw a grid like the one shown in Figure 3-2(a).** It needs to have as many columns as the longer number has digits, plus one more, and as many rows as the shorter number has digits plus one more. Here, it's a 4 by 2 grid – give yourself plenty of room.

2. **Write the short number in the bottom-left square.**

3. **Split the long number up into the values of its digits (for example: 200, 30 and 4) and put each in a new column.** Now it should look like Figure 3-2(b).

4. **In each square of the grid, work out the number at the top multiplied by the number at the left-hand end, as shown in Figure 3-2(c).** Here, that's 200 × 7 = 1,400; 30 × 7 = 210; 4 × 7 = 28.

5. **Add up all the numbers you've worked out to get your answer, as shown in Figure 3-2(d).** So, 1,400 + 210 + 28 = 1,638.

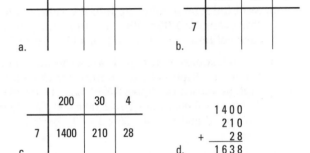

Figure 3-2: Multiplying a single number using a grid.

Working out long multiplication

The times tables grid works just the same with bigger numbers. If (heaven forbid) you had to multiply 123456789 × 987654321, you could do so with the grid. It might take you forever, but you could do it.

Luckily, nobody's going to ask you to do that kind of sum. At the very worst, you can expect a four-digit number multiplied by a two-digit number, say, 1,492 × 24.

Here's the full recipe:

1. **Draw a grid like the one in Figure 3-3(a).** It needs to have as many columns as the longer number has digits, plus one more, and as many rows as the shorter number has digits, plus one more. Here, it's a 5 by 3 grid – give yourself plenty of room.

2. **Split the shorter number into the values of its digits (20 and 4) and put them in the rows.**

3. **Split the long number up into the values of its digits (for example: 1,000, 400, 90 and 2) and put each in a new column.** Now it should look like Figure 3-3(b).

4. **In each square of the grid, work out the number at the top multiplied by the number at the left-hand end, as shown in Figure 3-3(c).** So, 1,000 × 20 = 20,000; 1,000 × 4 = 4,000; 400 × 20 = 8,000; 400 × 4 = 1,600; 90 × 20 = 1,800; 90 × 4 = 360; 2 × 20 = 40; 2 × 4 = 8. Phew!

5. **Add up all the numbers you've worked out to get your answer!** That's 20,000 + 4,000 + 8,000 + 1,600 + 1,800 + 360 + 40 + 8 = 35,808, as shown in Figure 3-3(d).

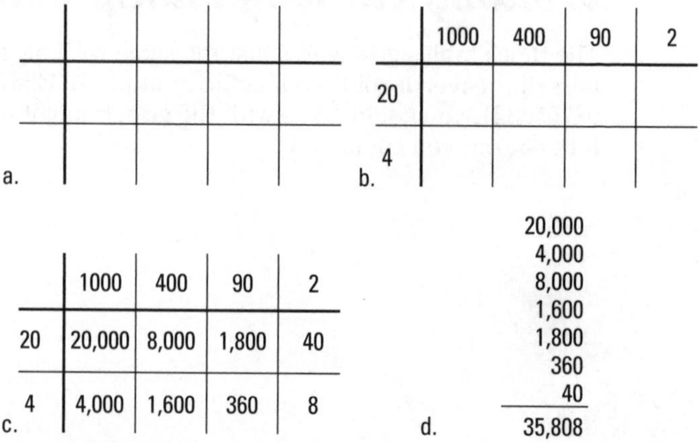

Figure 3-3: Multiplying big numbers using a grid.

Attempting some multiplication questions

1. Work out:

(a) 2×7 (b) 3×6 (c) 5×3 (d) 4×3

(e) 3×8 (f) 2×4 (g) 1×5 (h) 3×7

2. Work out:

(a) 4×9 (b) 8×7 (c) 9×6 (d) 5×8

(e) 7×9 (f) 6×8 (g) 7×6 (h) 8×8

3. Work out:

(a) 15×10 (b) 20×100 (c) 95×1000

4. What is...

(a) 135×3 (b) 942×4 (c) 263×5 (d) 1234×6

(e) 153×7 (f) 999×8 (g) 524×9 (h) 294×7

5. Work out:

(a) 401×5 (b) 937×5 (c) 512×2 (d) 935×4

(e) 651×3 (f) 862×7 (g) 328×3 (h) 941×9

6. Work out:

(a) 123×12 (b) 524×24 (c) 732×15

(d) 80×46 (e) 37×102 (f) 912×42

The answers are provided at the end of the chapter.

Splitting Things Up: Division

When you share anything out fairly, you're doing a division sum. If you have 20 cakes to share between four people, each person gets five cakes; that's the same as saying $20 \div 4 = 5$.

Dividing isn't always that straightforward, though: you can't always split a number up fairly. If you've got nine cakes to split between four people, everyone gets two – but you have one cake left over. (Don't worry, I'll eat that one later.) The two – what everyone gets – is called the *quotient*, and the 1 left over is the *remainder*. Here's how you find the quotient and remainder using the times tables grid shown in Figure 3-1 to work out $75 \div 9$:

1. **Look at the column beginning with the number after the ÷ sign.**

2. **Find the biggest number in the column that's smaller than the number before the ÷ sign.** Here, it's 72.

3. **Work out the remainder, which is the difference between the number you just found and the number before the ÷ sign.** Here, that's $75 - 72 = 3$; 3 is the remainder.

4. **Work out the quotient, which is the number at the start of the row of the number you found.** Here, it's 8.

So, $75 \div 9 = 8$, with remainder 3.

To be able to do more with the remainder you first need to know about fractions or decimals. Take a look at Chapters 5 and 6 if you want to get started on those.

Working out short division

Working out division is all well and good if your number is in the times tables grid – but what if it's bigger than that? For bigger numbers you need short division – or the *bus stop* method.

You can do division in lots of different ways. I've only got room for one method here, though, so I'm going with the bus stop. As always, whichever way works for you is good – so if you know a method that works better for you, don't be afraid to use it!

The idea of the bus stop method is to work from left to right through the number, dividing a bit at a time and passing the remainder on to the next number to split up. It sounds a bit confusing, and it takes a few goes to get the hang of it – but you can do it! Here's how you'd work out 711 ÷ 9:

1. **Draw out a bus stop like the one shown in Figure 3-4(a), with the number you're splitting up under the shelter and the other number outside.** (Note: that means the number is the other way round from how it's presented in the sum.) Give yourself plenty of room between the numbers and above the bus stop to write in.

2. **Look at the first number under the bus stop and divide it by the number in front.** Here, 7 ÷ 9 gives a quotient of 0 with remainder 7.

3. **Write the quotient above the number you just dealt with, above the bus stop; write the remainder just in front of the next digit to the right.** It should look like Figure 3-4(b). Now the 1 looks like 71.

4. **Look at the next number under the bus stop (if there is one); here, it's 71.** Divide this by the number in front. So, 71 ÷ 9 = 7, remainder 8.

5. **Write the quotient above the number, outside of the bus stop, and put the remainder in front of the digit to the right (which will now be 81).**

6. **Repeat Steps 4 and 5 until you run out of numbers!** The next step is 81 ÷ 9 = 9, and that's the dividing done.

7. **Check out your answer on top of the bus stop.** In this case, it should be 79. Your working out should look like that shown in Figure 3-4(c).

$$\begin{array}{r} 0 \\ 9 \overline{)7 \quad 1 \quad 1} \end{array}$$
a.

$$\begin{array}{r} 0 \\ 9 \overline{)7 \quad 71 \quad 1} \end{array}$$
b.

$$\begin{array}{r} 0 \quad 7 \quad 9 \\ 9 \overline{)7 \quad 71 \quad 81} \end{array}$$
c.

Figure 3-4: Working out the steps in short division.

Cancelling down

Once you get the hang of dividing by single-digit numbers, the logical next step is dividing by two-digit numbers. I show you two ways to do this: cancelling down, in this section, and long division proper in the next one.

Cancelling down is a really powerful method for making division sums easier when you can split the number you're dividing by into smaller numbers. Here's how to do 136,488 ÷ 24, the easy way:

1. **Look at the number you're dividing by and find two numbers that multiply together to make it.** Here, you might pick 4 and 6 (because 4 × 6 = 24) or 3 and 8 (3 × 8 = 24 as well). I'll go with 4 and 6.

2. **Divide the big number by the first of your numbers in Step 1.** You get 136,488 ÷ 4 = 34,122.

3. **Divide your answer by the second number.** You get 34,122 ÷ 6 = 5687, which is your answer.

This method only works when you can split your number up neatly! If you want to divide by a number like 37, which isn't in the times tables, you have to do it with long division.

Getting to the bottom of long division

Take a deep breath. Long division is one of those topics that everyone thinks is really hard and scary, but it's really no harder than short division if you do it right. Long division involves a three-step process, and all three steps are easy:

✔ Write the times table for the number you're dividing by.

✔ Draw out the bus stop you used for short division – but with more room.

✔ Divide each number in turn, like for short division, using the times table grid shown in Figure 3-1.

In this section, I show you how to work out 3,515 ÷ 19.

Working out the times table

The first step is to work out the times table for the number you're dividing by. Here's how you do it for 19:

1. **Write the numbers 1 to 10 in a column on the left.**

2. **Write down your number next to the 1.**

3. **Add your number on to the last number you wrote down and write it underneath.**

4. **Repeat Step 3 until you get up to the tenth number.** If you've done it right, you'll recognise 10 × your number; here, it should be 190. Take a look at Figure 3-5(a) to make sure you've done it right.

Doing the sum

Now it's time for the dividing! The most important thing here is to leave plenty of space between the numbers under the bus stop – you need enough room to write at least two more numbers between each of them.

1. **Draw a bus stop like the one in Figure 3-5(b).** Put the number you're splitting up under the shelter and the other number outside.

2. **Look at the first number under the bus stop and divide it by the number in front.** Here, 3 ÷ 19 gives a quotient of 0, remainder 3.

3. **Write the quotient above the number you just dealt with, above the bus stop; write the remainder just in front of the next digit to the right.** It should look like Figure 3-5(c). Now the 5 looks like 35.

4. **Look at the next number under the bus stop (if there is one); here, it's 35.** Divide this by the number in front, by looking at the times table you wrote out: 35 ÷ 19 = 1, remainder 16.

5. **Write the quotient above the number, outside of the bus stop, and put the remainder in front of the digit to the right (which will now be 161).**

6. **Repeat Steps 4 and 5 until you run out of numbers.** The next steps will be 161 ÷ 19 = 8, remainder 9; then 95 ÷ 19 = 5. Now you're done.

7. **Check out your answer, which is the number on top of the bus stop.** In this case, it's 185. Your workings out should look like those in Figure 3-5(d).

```
        1    19
        2    38
        3    57
        4    76
        5    95
        6    114
        7    133
        8    152
        9    171
   a.   10   190          b.  19 ) 3   5   1   5
```

```
            0                         0   1   8   5
   c.  19 ) 3  35   1   5      d.  19 ) 3  35  161  95
```

Figure 3-5: Working through the steps in long division.

Attempting some division questions

1. Work out:

(a) 10 ÷ 2   (b) 6 ÷ 2   (c) 9 ÷ 3   (d) 35 ÷ 5

(e) 16 ÷ 2   (f) 12 ÷ 3   (g) 24 ÷ 6   (h) 25 ÷ 5

2. Work out:

(a) 54 ÷ 6   (b) 18 ÷ 9   (c) 36 ÷ 4   (d) 72 ÷ 9

(e) 64 ÷ 8   (f) 32 ÷ 8   (g) 27 ÷ 9   (h) 56 ÷ 8

3. Work out:

(a) 800 ÷ 8   (b) 40 ÷ 4   (c) 9,000 ÷ 9   (d) 20,000 ÷ 2

(e) 800 ÷ 2   (f) 40 ÷ 2   (g) 9,000 ÷ 3   (h) 20,000 ÷ 5

4. Work out:

(a) 978 ÷ 2   (b) 978 ÷ 3   (c) 776 ÷ 4   (d) 215 ÷ 5

(e) 978 ÷ 6   (f) 1,120 ÷ 7   (g) 776 ÷ 8   (h) 891 ÷ 9

5. Work out by cancelling:

(a) 600 ÷ 12   (b) 480 ÷ 24   (c) 162 ÷ 18   (d) 876 ÷12

(e) 720 ÷ 36   (f) 1,120 ÷ 14   (g) 2,250 ÷ 25   (h) 1,080 ÷ 36

6. Work out by long division (or using whichever method you prefer):

(a) 9,690 ÷ 17 (b) 3,874 ÷ 26

(c) 3,813 ÷ 31 (d) 11,298 ÷ 42

The answers are provided at the end of the chapter.

Working through Review Questions

1. (a) 2 × 8   (b) 3 × 6   (c) 3 × 2   (d) 4 × 4

(e) 2 × 7   (f) 6 × 2   (g) 7 × 1   (h) 3 × 5

2. (a) 7 × 7   (b) 4 × 9   (c) 7 × 4   (d) 7 × 5

(e) 6 × 9   (f) 9 × 9   (g) 7 × 6   (h) 8 × 4

3. Work out:

(a) 4×10　　(b) 90×10　　(c) 9×100　　(d) 45×1000

4. Work out:

(a) 400×7　　(b) 30×5　　(c) $3{,}000 \times 4$　　(d) $80{,}000 \times 2$

5. Work out:

(a) 370×8　　(b) 780×5　　(c) 296×3　　(d) 729×3

(e) 877×7　　(f) 937×7　　(g) 931×8　　(h) 158×6

6. Work out:

(a) $3{,}452 \times 5$　　(b) $3{,}884 \times 4$　　(c) $8{,}429 \times 4$

(d) $8{,}376 \times 6$　　(e) $3{,}999 \times 9$　　(f) $3{,}224 \times 8$

7. Work out:

(a) 328×17　　(b) 598×13　　(c) 927×15

(d) 861×25　　(e) 511×14　　(f) 151×15

8. Work out:

(a) $24 \div 4$　　(b) $24 \div 3$　　(c) $14 \div 2$　　(d) $25 \div 5$

(e) $36 \div 6$　　(f) $21 \div 3$　　(g) $16 \div 4$　　(h) $35 \div 5$

9. Work out:

(a) $42 \div 7$　　(b) $72 \div 8$　　(c) $81 \div 9$　　(d) $56 \div 7$

(e) $54 \div 6$　　(f) $49 \div 7$　　(g) $64 \div 8$　　(h) $48 \div 6$

10. Work out:

(a) $960 \div 10$　　(b) $400 \div 100$　　(c) $15{,}120 \div 10$　　(d) $1{,}200 \div 10$

11. Work out:

(a) $4,200 \div 6$ (b) $16,000 \div 4$ (c) $2,500 \div 5$ (d) $560 \div 8$

(e) $900 \div 30$ (f) $490 \div 70$ (g) $16,000 \div 400$ (h) $20,000 \div 50$

12. Work out:

(a) $68 \div 2$ (b) $78 \div 2$ (c) $364 \div 4$ (d) $365 \div 5$

(e) $686 \div 7$ (f) $856 \div 8$ (g) $396 \div 9$ (h) $595 \div 7$

13. Work out:

(a) $4,764 \div 2$ (b) $4,764 \div 3$ (c) $4,764 \div 4$ (d) $2,470 \div 5$

(e) $2,472 \div 6$ (f) $2,065 \div 7$ (g) $1,960 \div 8$ (h) $2,205 \div 9$

14. Work out by cancelling:

(a) $1,392 \div 12$ (b) $1,056 \div 16$ (c) $1,872 \div 18$ (d) $3,168 \div 24$

15. Work out:

(a) $3,196 \div 17$ (b) $3,572 \div 19$ (c) $1,496 \div 17$ (d) $923 \div 13$

Checking Your Answers

Make sure you take plenty of time to work through the questions in this chapter. You have a lot to cover – but once you know how to do long division, you'll be able to show off to all your friends.

Multiplying questions

1. (a) 14 (b)18 (c) 15 (d)12

(e) 24 (f) 8 (g) 5 (h) 21

2. (a) 36   (b) 56   (c) 54   (d) 40

  (e) 63   (f) 48   (g) 42   (h) 64

3. (a) 150   (b) 2,000   (c) 95,000

4. (a) 405   (b) 3,768   (c) 1,315   (d) 7,404

  (e) 1,071   (f) 7,992   (g) 4,716   (h) 2,058

5. (a) 2,005   (b) 4,685   (c) 1,024   (d) 3,740

  (e) 1,953   (f) 6,034   (g) 984   (h) 8,469

6. (a) 1,476   (b) 12,576 (c) 10,980

  (d) 3,680   (e) 3,774   (f) 38,304

Division questions

1. (a) 5   (b) 3   (c) 3   (d) 7

  (e) 8   (f) 4   (g) 4   (h) 5

2. (a) 9   (b) 2   (c) 9   (d) 8

  (e) 8   (f) 4   (g) 3   (h) 7

3. (a) 100   (b) 10   (c) 1,000   (d) 10,000

  (e) 400   (f) 20   (g) 3,000   (h) 4,000

4. (a) 489   (b) 326   (c) 194   (d) 43

  (e) 163   (f) 160   (g) 97   (h) 99

5. (a) $600 \div 12 = 300 \div 6 = 150 \div 3 = 50$

   (b) $480 \div 24 = 240 \div 12 = 120 \div 6 = 60 \div 3 = 20$

   (c) $162 \div 18 = 81 \div 9 = 9$

   (d) $876 \div 12 = 438 \div 6 = 219 \div 3 = 73$

   (e) $720 \div 36 = 360 \div 18 = 180 \div 9 = 20$

   (f) $1{,}120 \div 14 = 560 \div 7 = 80$

   (g) $2{,}250 \div 25 = 450 \div 5 = 90$

   (h) $1{,}080 \div 36 = 540 \div 18 = 270 \div 9 = 30$

6. (a) 570   (b) 149   (c) 123   (d) 269

Review questions

1. (a) 16   (b) 18   (c) 6   (d) 16

   (e) 14   (f) 12   (g) 7   (h) 15

2. (a) 49   (b) 36   (c) 28   (d) 35

   (e) 54   (f) 81   (g) 42   (h) 32

3. (a) 40   (b) 900   (c) 900   (d) 45,000

4. (a) 2,800   (b) 150   (c) 12,000   (d) 160,000

5. (a) 2,960   (b) 3,900   (c) 888   (d) 2,187

   (e) 6,139   (f) 6,559   (g) 7,448   (h) 948

6. (a) 17,260   (b) 15,536   (c) 33,716

    (d) 50,256   (e) 35,991   (f) 25,792

7. (a) 5,576   (b) 7,774   (c) 13,905

    (d) 21,525   (e) 7,154   (f) 2,265

8. (a) 6   (b) 8   (c) 7   (d) 5

    (e) 6   (f) 7   (g) 4   (h) 7

9. (a) 6   (b) 9   (c) 9   (d) 8

    (e) 9   (f) 7   (g) 8   (h) 8

10. (a) 96   (b) 4   (c) 1,512   (d) 120

11. (a) 700   (b) 4,000   (c) 500   (d) 70

    (e) 30   (f) 7   (g) 40   (h) 400

12. (a) 34   (b) 39   (c) 91   (d) 73

    (e) 98   (f) 107   (g) 44   (h) 85

13. (a) 2,382   (b) 1,588   (c) 1,191   (d) 494

    (e) 412   (f) 295   (g) 220   (h) 245

14. (a) $1,392 \div 12 = 696 \div 6 = 348 \div 3 = 116$

    (b) $1,056 \div 16 = 528 \div 8 = 264 \div 4 = 132 \div 2 = 66$

    (c) $1,872 \div 18 = 936 \div 9 = 312 \div 3 = 104$

    (d) $3,168 \div 24 = 1584 \div 12 = 792 \div 6 = 396 \div 3 = 132$

15. (a) 188   (b) 188   (c) 88   (d) 71

Chapter 4

Are We Nearly There Yet? Estimating and Rounding

● ●

In This Chapter

▶ Rounding numbers up or down

▶ Getting a rough answer

▶ Following BIDMAS

● ●

*T*his may go against everything your maths teacher ever taught you, but here goes: sometimes it's okay to come up with a rough answer. For instance, if you talk about a posh house being for sale for a million pounds, it doesn't matter too much if the price is really £950,000 or £1,032,795 or anything else around that number. If you call it a million quid, you're close enough – and it's a bit easier to understand than either of the more accurate numbers.

Here's another example: you get to the supermarket checkout and find you've lost your wallet but, luckily, find a £10 note in your pocket. You'll probably want to know if the groceries in your basket cost much less than £10, much more than £10 or about £10 as quickly as possible. Rather than getting your pen and paper out, you could mentally run through the contents, saying to yourself, 'that costs about £2, those things together cost £3. . .' and so on. If the total turns out to be close to £10, then you'll probably get the pen and paper out (or, more likely, explain everything to the cashier, and possibly leave a loaf of bread behind).

This chapter is all about getting quick and dirty answers – ones that aren't exactly right, but good enough to be going on with. I take you through rounding off, finding approximate answers and – on a note that's only slightly related – dealing with formulas.

Rounding Numbers Off

Rounding a number means taking a precise number that's tricky to deal with – such as 2.71828 – and finding a less accurate number that's (a) easier to work with and (b) close enough to be acceptable. Depending on how accurate you needed to be, you might round 2.71828 to:

- ✔ 3, if you wanted the nearest whole number
- ✔ 2.7, if you wanted it to one decimal place
- ✔ 2.72 if you wanted it to two decimal places

and so on. In this section, I show you how to get to those numbers.

Naming the nearest number

Most of the rounding you'll need to do involves finding the nearest whole number, the nearest 10 or the nearest 100 to a given number.

Imagine, for a moment, that you have £13.86 in your pocket. That amount of money is between £13 and £14, obviously – but it's closer to £14 than it is to £13. *To the nearest pound*, it would be £14. It's also between £10 and £20 – but closer to £10 than it is to £20. So, *to the nearest £10*, it would be £10.

To round your value off to the nearest whole number, the nearest 10, nearest 100 or whatever:

1. **Find the numbers above and below your value that have the right precision.** For instance, to round 713 to the nearest 100, you say, 'it's between 700 and 800'.

2. **Decide whether your value is above or below halfway.** Here, halfway would be 750, and 713 is below that.

3. **Give your answer.** If it's below halfway, your answer is the lower number from Step 1. If it's above halfway, it's the bigger number from Step 1.

What happens if the number you want to round is exactly halfway, though? Good question. There's a rule for that situation: *if your number is exactly halfway between your lower and upper bound, you round up*. So, to work out 1,500 to the nearest thousand, you say, 'it's between 1,000 and 2,000; it's exactly halfway, so I round *up* to 2,000.'

Dealing with decimal places

Working out numbers to one, two or any other number of decimal places is no harder than rounding to the nearest whole number. Here's how it works – I'm going to round 3.1415927 to three decimal places:

1. **Count right from the dot the number of decimal places you want to round to, and draw a vertical line after that digit.** Here, you draw a line after the second 1, making 3.141|5927.

2. **If the digit *after* the line is 4 or smaller, just chop off all the numbers to the right.** That's your answer.

3. **If the digit after the line is 5 or higher, add 1 to the digit *before* the line (and carry – see Chapter 3 for the lowdown on carrying – if you need to), – then chop off all the numbers to the right of the line.** Here, I'd bump the second 1 up to a 2 and ignore everything after it, – getting 3.142. That's the answer.

Attempting rounding questions

1. To the nearest 10, what is?

(a) 19   (b) 34   (c) 59   (d) 437

2. To the nearest 100, what is?

(a) 1,563   (b) 13,214   (c) 522   (d) 95

3. To the nearest 1,000, what is?

(a) 6,325   (b) 2,445   (c) 1,852   (d) 9,191

4. To the nearest 1,000, what is?

(a) 1,521   (b) 726,135   (c) 916,231   (d) 13,152

5. To one decimal place, what is?

(a) 6.55   (b) 5.45   (c) 9.165   (d) 45.52

Answering approximately

Giving an answer that's good enough is sometimes good enough! I'm a big fan of rough answers because they can tell you the kind of answer you're looking for before you even start working, so you can tell straight away if you've gone badly wrong. For instance, if I have to work out 2.05×380, I would probably say, 'that's about the same as 2×400, so I expect an answer of about 800'. When I work it out and get 779, I can nod my head and say, 'that looks reasonable'; if I hadn't done the rough sum first, I'd have no idea whether I was in the right ballpark.

As well as giving you some confidence in the answer to any sum, some tests ask you explicitly to give approximate answers. There's not really a recipe for this kind of question, because each question has different ideas about how accurately you have to round things, but the idea is generally:

✔ Round off each number you're given to the accuracy the question asks for.

✔ Work out the sum using the rounded numbers.

Rounding involves really no more than that!

Attempting some approximation questions

1. Suzie approximates her shopping bill by rounding all of the amounts to the nearest pound. What is her approximation of:

(a) A litre of milk that costs £1.20?

(b) A pizza that costs £4.90?

(c) A bottle of wine that costs £8.49?

(d) Her shopping bill for this short and slightly sad list of items?

2. Swar approximates his earnings by rounding his number of hours worked to the nearest ten hours, and his hourly rate to the nearest pound. How does he approximate:

(a) Week 1, in which he worked 32 hours at £8.65 per hour?

(b) Week 2, in which he worked 46 hours at £7.95 per hour?

(c) Week 3, in which he worked 22 hours at £9.15 per hour?

(d) His total wages for the three weeks?

The answers are provided at the end of the chapter.

Bracing Yourself for BIDMAS

You may remember your maths teacher, at some point in the past, explaining why something had gone wrong 'because of BIDMAS' – or PEMDAS or BODMAS or some other made-up word that was supposed to be an explanation.

BIDMAS is a very useful tool, as long as you know how to use it. It stands for brackets, indices, division and multiplication, add and subtract, and tells you what order to do a complicated sum in.

In this section, I show you how BIDMAS works in more detail, and show you how to use *formulas* to work out some more tricky sums.

Investigating indices

Chapter 2 covers adding and subtracting and Chapter 3 deals with multiplication and division. Which leaves one more *operation* you need to know about: *indices* or *powers*.

An index is simply a little number written above and right of another number, like this: 15^2. All it means is, multiply the base number (here, 15) by itself the number of times in the index. So, 15^2 is the same as $15 \times 15 = 225$. Similarly, $2^5 = 2 \times 2 \times 2 \times 2 \times 2 = 32$.

If you go on to do more maths, you'll get to learn about negative and fractional indices, which are great fun but slightly more involved. For basic maths, you just need to know that the little number up top is how many times you multiply the number by itself.

Doing things in order

Maths is a very precise language – and one that generally avoids saying anything it doesn't need to. Part of the grammar of maths is the *order of operations*, which tells you which sums to do first. Here's why the order is important; look at this sum:

$15 + 8 \times 9 - 6 \div 2$

If you don't know in which order to do the operations, I reckon 24 different orders are possible. If you do the take away sum $(9 - 6)$ first, you get a different answer than if you do the times sum (8×9) first. Try it and see!

However, this sum (and any sum you're likely to see) only has one correct answer – and BIDMAS tells you how to get to it. Follow these steps:

1. **B.** If your sum has any *brackets* in it, work out their values first, using BIDMAS. Write the sum out again, with the brackets replaced by the numbers they worked out to. This sum doesn't have any brackets.

2. **I.** If your sum has any *indices* in it, work them out next. Write the sum again, with all of the index sums replaced by their values.

3. **DM.** If your sum has any dividing and/or multiplying in it, work these out from left to right. In this sum, you work out $8 \times 9 = 72$ and $6 \div 2 = 3$, then rewrite the sum as $15 + 72 - 3$.

4. **AS.** Now work out any adding and subtractions from left to right. So, $15 + 72 - 3 = 87 - 3 = 84$, which is the one correct answer!

Here's a more complicated sum: $(10 - 3 \times 2)^2 + 7 \times 3 - 5$.

You do the bracket first, which is $10 - 3 \times 2$; BIDMAS applies in this sum, too, so you work out $3 \times 2 = 6$ to get $10 - 6 = 4$. The value of the bracket is 4, making the sum:

$$4^2 + 7 \times 3 - 5$$

The next thing to deal with is the index, the little 2 above the 4. That means $4 \times 4 = 16$, giving you:

$$16 + 7 \times 3 - 5$$

Next it's the multiply: $7 \times 3 = 21$, leaving:

$$16 + 21 - 5$$

And finally you can work this out to be 32. Phew!

BIDMAS stands for Brackets, Indices, Divide and Multiply, Add and Subtract. If you do the sums in that order, you'll get them right!

Finding the formula

This book doesn't cover algebra, but basic maths does include a little bit of 'replacing letters with numbers'; that is, using formulas to work out a value.

You see formulas all over the place. Famous examples are $E = mc^2$, $A = \pi r^2$ and, of course, $D = S + X - N_c - 1$, the Beveridge–Longcope equation that stops the world from being destroyed by solar flares. (So far, so good.)

The point of these alphabetti spaghetti concoctions is that each letter represents a value, and if you replace the letters with the values given to you, you can work out another value. For instance, if you know the mass of an object (m) and the speed of light (c), you can use $E = mc^2$ to work out how much energy (E) the object has.

More often, you'll get a formula like $V = hx^2 \div 3$, where V is the volume of a pyramid in m^3, x is how wide it is in metres and h is its height in metres. Here's how you'd use it to find the volume of the Pyramid of Menkaure, the smallest of the famous pyramids in Giza, which is 100 metres wide and 69 metres tall:

1. **If you have two letters next to each other, or a number in front of a letter, write a × sign between them.** This turns the formula into $V = h \times x^2 \div 3$.

2. **Replace each letter with the number given in the formula.** Doing so makes $V = 69 \times 100^2 \div 3$.

3. **Work out the formula using BIDMAS!** Here, you do the power first, giving $69 \times 10,000 \div 3$, then the multiplication and division from left to right, giving $690,000 \div 3 = 230,000\text{m}^3$.

Attempting some BIDMAS questions

1. Work out:

(a) $3 \times 4 + 2$

(b) $3 \times (4 + 2)$

(c) $(3 + 4) \times (5 + 6)$

(d) $3 + 4 \times 5 + 6$

(e) $(13 - 9) \div (9 - 7)$

(f) $13 - 9 \div 9 - 7$

(g) $12 - 4 + 7 - 3$

(h) $12 - (4 + 7) - 3$

2. The stopping distance in centimetres for a car travelling at v miles per hour (mph) is given by the formula: $30v + 1.5v^2$. What is the stopping distance for a car travelling at:

(a) 20mph?   (b) 50mph?   (c) 70mph?   (d) 100mph?

3. The volume of water in a particular style of fishpond in litres is given by the formula $2\,r^2h \,/\, 3000$, where r is the radius of the pond in centimetres and h is the depth of water in centimetres. How much water is in the pond if:

(a) The radius is 300cm and the depth is 20cm?

(b) The radius is 400cm and the depth is 30cm?

(c) The radius is 100cm and the depth is 30cm?

(d) The radius is 200cm and the depth is 9cm?

4. A sales rep gets a basic salary of £2,000 a month, plus 10 per cent commission on sales. Which of the following is the formula she'd use to work out her pay if she made £15,000 of sales in one month?

A: $2,000 + 100 \times 10 \div 15,000$

B: $2,000 + 15,000 \times 100 \div 10$

C: $2,000 \times 15,000 \times 10 \div 100$

D: $2,000 + 15,000 \times 10 \div 100$

The answers are provided at the end of the chapter.

Working through review questions

1. Round the numbers below to the nearest ten:

(a) 292　　　(b) 739　　　(c) 4,314　　　(d) 1,678

2. Round the numbers below to the nearest 100:

(a) 769　　　(b) 437　　　(c) 9,275　　　(d) 6,987

3. Round the numbers below to the nearest 1,000:

(a) 33,143　　(b) 74,854　　(c) 29,696　　(d) 83,499

4. Round the numbers below to the nearest whole number:

(a) 2.63　　　(b) 0.62　　　(c) 59.313　　　(d) 228.77

5. To one decimal place, what is:

(a) 407.816?　(b) 403.556?　(c) 302.517?　(d) 74.199?

6. What is:

(a) 14.45 to the nearest ten?

(b) 14.45 to the nearest whole number?

(c) 14.45 to one decimal place?

7. Alice is working out her budget for the month. She decides to round all of her bills to the nearest £10. What is her estimate for:

(a) Her electricity and gas bill of £127.50?

(b) Her water bill of £97?

(c) Her phone and Internet bill of £44.50?

(d) Her four grocery trips, each of which costs £53 (she works out the total before she rounds)?

(e) Her petrol bill (it costs £63 to fill up her car, and she expects to do that three times)?

(f) Her total expenses for the month, – including her rent of £600?

8. Work out:

(a) $(45 - 5) \times (6 + 3 \div 3)$

(b) $45 - 5 \times 7 + 3 \div 3$

(c) $45 - 5 \times (6 + 3) \div 3$

(d) $(45 - 5) \times (6 + 3) \div 3$

9. The power, P, in an electric circuit (in watts) is given by the formula $P = V^2 \div R$, where V is the potential difference (in volts) and R is the resistance (in ohms). What is the power if:

(a) The resistance is 10 ohms and the potential difference is 20 volts?

(b) The resistance is 20 ohms and the potential difference is 10 volts?

(c) The resistance is 30 ohms and the potential difference is 30 volts?

(d) The resistance is 5 ohms and the potential difference is 10 volts.

Resistance is futile! Don't worry too much about what resistance, potential difference and current mean – you can do the sums without understanding electronics!

The answers are provided at the end of the chapter.

Checking Your Answers

Estimating and using formulas are two of the most important skills in basic maths, so spending some time on these topics to make sure you understand them is certainly worth it.

Rounding questions

1. (a) 20   (b) 30   (c) 60   (d) 440

2. (a) 1,600 (b) 13,200   (c) 500   (d) 100

3. (a) 6,000 (b) 2,000   (c) 2,000   (d) 9,000

4. (a) 1,500 (b) 726,000 (c) 916,000 (d) 13,000

5. (a) 6.6   (b) 5.5   (c) 9.2   (d) 45.5

Approximation questions

1. (a) £1   (b) £5   (c) £8   (d) £14 (the actual total is £14.59)

2. (a) 30×9 = £270   (b) 50×8 = £400   (c) 20×9 = £180   (d) £850 (the actual total is £843.80)

BIDMAS questions

1. (a) $3 \times 4 + 2 = 12 + 2 = 14$

   (b) $3 \times (4 + 2) = 3 \times 6 = 18$

   (c) $(3 + 4) \times (5 + 6) = 7 \times 11 = 77$

   (d) $3 + 4 \times 5 + 6 = 3 + 20 + 6 = 29$

   (e) $(13 - 9) \div (9 - 7) = 4 \div 2 = 2$

   (f) $13 - 9 \div 9 - 7 = 13 - 1 - 7 = 5$

   (g) $12 - 4 + 7 - 3 = 8 + 4 = 8 + 7 - 3 = 15 - 3 = 12$

   (h) $12 - (4 + 7) - 3 = 12 - 11 - 3 = 1 - 3 = -2$'

2. (a) $30 \times 20 + 1.5 \times 20^2 = 30 \times 20 + 1.5 \times 400 = 600 + 600 = 1200$cm or 12 m

   (b) $30 \times 50 + 1.5 \times 50^2 = 30 \times 50 + 1.5 \times 2500 = 1500 + 3750 = 5250$cm or 52.5 m

   (c) $30 \times 70 + 1.5 \times 70^2 = 30 \times 70 + 1.5 \times 4900 = 2100 + 7350 = 9450$cm or 94.5 m

   (d) $30 \times 100 + 1.5 \times 100^2 = 30 \times 100 + 1.5 \times 10{,}000 = 3000 + 15{,}000 = 18{,}000$cm or 180 m

3. (a) $2 \times 300^2 \times 20 \div 3{,}000 = 2 \times 90{,}000 \times 20 \div 3{,}000 = 3{,}600{,}000 \div 3{,}000 = 1{,}200$ litres

   (b) $2 \times 400^2 \times 30 \div 3{,}000 = 2 \times 160{,}000 \times 30 \div 3{,}000 = 9{,}600{,}000 \div 3{,}000 = 3{,}200$ litres

   (c) $2 \times 100^2 \times 30 \div 3{,}000 = 2 \times 10{,}000 \times 30 \div 3{,}000 = 600{,}000 \div 3{,}000 = 200$ litres

   (d) $2 \times 200^2 \times 9 \div 3{,}000 = 2 \times 40{,}000 \times 9 \div 3{,}000 = 720{,}000 \div 3{,}000 = 240$ litres

4. D: $2000 + 15{,}000 \times 10 \div 100$. That's the basic salary, plus 10 per cent of 15,000.

Review questions

1. (a) 290   (b) 740   (c) 4,310   (d) 1,680

2. (a) 800   (b) 400   (c) 9,300   (d) 7,000

3. (a) 33,000   (b) 75,000   (c) 30,000   (d) 83,000

4. (a) 3   (b) 1   (c) 59   (d) 229

5. (a) 407.8   (b) 403.6   (c) 302.5   (d) 74.2

6. (a) 10   (b) 14   (c) 14.5

7. (a) £130   (b) £100   (c) £40

   (d) $53 @ 4 = 212$, or £210 to the nearest £10   (e) $63 \times 3 = 189$, or £190 to the nearest £10   (f) $130 + 100 + 40 + 210 + 190 + 600 = £1,270$

8. (a) $(45 - 5) \times (6 + 3 \div 3) = 40 \times 7 = 280$

   (b) $45 - 5 \times 7 + 3 \div 3 = 45 - 35 + 1 = 11$

   (c) $45 - 5 \times (6 + 3) \div 3 = 45 - 5 \times 9 \div 3 = 45 - 45 \div 3 = 45 - 15 = 30$

   (d) $(45 - 5) \times (6 + 3) \div 3 = 40 \times 9 \div 3 = 360 \div 3 = 120$

9. (a) $20 \times 20 \div 10 = 40$

   (b) $10 \times 10 \div 20 = 5$

   (c) $30 \times 30 \div 30 = 30$

   (d) $10 \times 10 \div 5 = 20$

Part II
Working with Parts of the Whole

In this part . . .

If simple arithmetic was all there was to maths, it would be a really easy – but also really dull – subject. In Part II, I take it a bit further and give you practice questions on fractions, decimals, percentages and ratios – all of which are ways of talking about things that aren't whole numbers.

Don't panic, though: I'm a gentle soul, and I show you straightforward ways to deal with what may look (at first) to be hard sums.

Chapter 5

Facing Fractions without Fear

● ●

In This Chapter

▶ Getting to grips with fractions

▶ Spotting the difference

▶ Dealing with fractions like pieces of cake

● ●

*F*irst up, take a deep breath: fractions are usually where maths starts to get hard for students, and it's easy to become anxious or panicky about them. Here are three important things to remember:

✔ You can do an awful lot of maths without using fractions (although fractions can make some things easier).

✔ You only need to remember a handful of things about fractions.

✔ Once you have a handle on fractions, you don't have to run away from them ever again!

In this chapter, I run you through what a fraction is and how to turn it into a picture and back again; I show you how to change the way you describe a fraction without changing its value; and I demonstrate how to apply that knowledge to sums. Working with fractions is much easier than it looks. Honestly!

Recognising Fractions

You've seen fractions before, right? They look like this: ⅚ or like this: $\frac{5}{8}$. Two numbers are presented there: the five (on the top, so I call it the . . . *top*) and the seven, which is on the *bottom*, so it's called . . . well, go on, guess.

In some maths books, you'll see the word *numerator*, which just means the top, and *denominator*, which means the bottom. I much prefer top and bottom.

You can think of a fraction as describing how much of a cake is left. The bottom number is how many slices were in the whole cake – so for ⅚, the cake would have had eight slices to begin with, as shown in Figure 5-1; the top says how many slices are left of the cake – in this case, five.

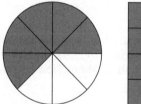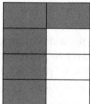

Figure 5-1: Two different ways of drawing $\frac{5}{8}$. Each drawing contains eight equal slices and five of them are coloured in.

Recognising names of fractions

You may need to be able to read the names of fractions and translate them into 'one number over another' form. To do so, you convert the fraction into an *ordinal number*. This number is what you'd use to describe where someone finished in a race, such as 'third' or 'fifteenth'. So, for example:

1. **If you see the word 'half', mentally replace it with 'second'.** If you see the word 'quarter', replace it with 'fourth'.

2. **If the fraction is a number followed by an ordinal number, the first number goes on top and the second on the bottom.** Three-eighths becomes ⅜; three-quarters becomes ¾.

3. **If the fraction is simply 'a' or 'an' followed by an ordinal number, put a 1 on top and the ordinal number on the bottom.** A fifth is ⅕ and a half is ½.

Finding the right fraction

If you have a picture – usually a circle or a rectangle – and you want to find out what fraction a part of it represents, follow these steps:

1. **Work out how many slices there were in the original shape.** You may need to draw extra lines in if you can't work it out straight away – make sure all the slices are the same size! In Figure 5-1, you can see that there are eight slices altogether.

2. **Count how many slices were part of the picture you were given.** In Figure 5-1, five of the eight slices are shaded.

3. **Write the number from Step 2, an oblique line (/) and the number from Step 1.** For this picture, you get ⅝.

Drawing the right picture

Drawing a picture that represents a fraction is quite similar to the process described in the preceding section, just the other way around. Follow these steps:

1. **If you're drawing the picture, start by drawing a circle.**

2. **Split it up into as many parts as the number on the bottom of your fraction (if you're drawing $\frac{5}{8}$, you split the circle into four parts).** Make sure the parts are all the same size!

3. **Find the number on top of the fraction and shade in that many of the parts.** That's it!

If you can draw a fraction like this, you can just as easily spot the fraction in a line-up! All you need to do is look at the bottom number and find the picture with that many slices in the original shape, and then look at the top number to see how many slices ought to be shaded.

Attempting some recognition questions

1. Write the following fractions as numbers:

(a) Two-fifths   (b) Three-tenths   (c) One-half

(d) Nine-sixteenths   (e) Three-quarters (f) An eighth

2. In the figure below, what fractions do (a), (b), (c) and (d) represent?

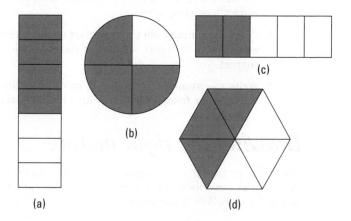

(a)   (d)

3. Draw the following fractions:

(a) ½   (b) ⅔   (c) ¾   (d) ⅛

(e) ⁹⁄₁₀   (f) ¼   (g) ⅝   (h) ⅚

Looking at top-heavy fractions and mixed numbers

Most of the fractions you'll see – at least in this book and at the level of the numeracy curriculum – will be smaller than one, or less than a whole cake. However, you may need to be able to deal with fractional numbers that are bigger than one.

You can write these monstrous fractions in two ways: as *top-heavy* fractions (with the top of the fraction bigger than the bottom) or as *mixed numbers*, which means a whole number followed by a fraction.

For instance, ³⁄₂ is a top-heavy fraction – the top is bigger than the bottom. You could also write it as 1½ because two half-cakes make a whole cake (the 1) and you have one half left over.

To convert a top-heavy fraction into a mixed number, follow these steps:

1. **Divide the top by the bottom (for example, $3 \div 2 = 1$, with remainder 1).**

2. **Write the *answer* from Step 1 fairly large.**

3. **If there was a *remainder* in Step 1, write it small after the answer from Step 2. Then draw an oblique line (/) and the number you originally divided by. You get 1½.**

To convert back, the steps are even easier. For instance, start with 2⅓:

1. **Multiply the big number by the bottom of the fraction ($2 \times 3 = 6$).**

2. **Add your answer to the top of the fraction ($1 + 6 = 7$).**

3. **Write your answer from Step 2 on top of the bottom of the fraction – here, you have $\frac{7}{3}$.**

4. Match the following fractions to 1) to 8) in the figure below:

(a) ¼   (b) ¾   (c) ⁶⁄₈   (d) ⅛

(e) ³⁄₇   (f) ⁵⁄₇   (g) ²⁄₄   (h) ½

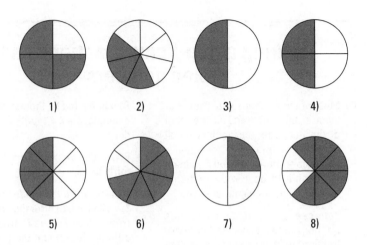

1)   2)   3)   4)

5)   6)   7)   8)

Keeping Things Fair

If you work through the questions in the previous exercise, you may notice something. (If you don't do the questions, or don't notice anything, that's totally okay, too).

The pictures you draw for 3(c) and 3(g) – ¾ and ⅚ – look very similar. In fact, if you take away the lines showing how big the slices are, you get two identical pictures.

In technical language, ¾ and ⅚ are *equivalent fractions*. Equivalent fractions are one of the most important things to understand about fractions because you use them nearly every time you do fraction sums, especially if you're adding or taking away.

In this section, I show you exactly how to tell when two fractions are equivalent, and how to change between the forms. I cover how to use this knowledge to solve sums later in the chapter!

Cancelling down

Two fractions are equivalent if you can multiply the top and bottom of one fraction by the same number and get the other fraction. So, ½ is equivalent to ³⁄₆ because you can multiply the top and bottom of the first fraction by 3 to get ³⁄₆ – and three-sixths of a cake is the same as half a cake!

Cancelling down a fraction (often called 'putting a fraction in its simplest terms' by well-meaning people who think, for no good reason, that that's clearer) just means taking the fraction you're given and finding the equivalent fraction with the smallest top (and bottom). Here are two ways of doing it:

1. **Find a number you can divide the top and the bottom by.** I usually look for 10s first, and then 2s, because they're easy to spot.

2. **Divide the top and bottom by the number you found in Step 1.**

3. **Repeat Steps 1 and 2 until you can't find anything more to divide by.** What you're left with is your answer.

Another method is one I fondly call 'fraction snap'. Before you start, you need to know how to factorise a number, which just means 'find numbers that multiply together to make it.' Here's how you do it with, for example, 48:

1. **Write down your number in a circle – give yourself plenty of room below it.**

2. **Find a number you can divide the number. Draw a line down to the left from your previous number and write this new number in a circle at the end of it. Two works here.**

3. **Divide your number from step 1 by the answer in step 2 – here, you'd get 24. Write that down at the end of a line coming down to the right from the original number.**

4. **Look for a number in your picture that doesn't have any lines coming out of it, and repeat steps 1-3 on this number. (If you can only divide the number by 1 and itself, leave it alone – it's a prime number).**

Here's how fraction snap works:

1. **Find something you can divide a number on the top by (for example, you can divide 12 by 2) and divide by it (here, you'd get 6).**

2. **Rewrite the top, replacing the number you just looked at with the two you came up with in step 1. In this case, the top would change from 12 to 2×6.**

3. **Do the same thing with the bottom – you might change 36 into 2 × 18.**

4. **If you have a pair of numbers, one on the top and one on the bottom, you can circle them both, say 'snap!' and get rid of them.**

5. **Repeat steps 1-4 until you can't go any further!**

6. **If you cross out the last number on top, replace it with a 1. If you cross out the last number on the bottom, it's not a fraction any more, and you're just left with the top!**

Figure 5-2 shows this process in action.

$$\frac{12}{36} = \frac{2 \times 2 \times 3}{2 \times 2 \times 3 \times 3}$$

$$\frac{12}{36} = \frac{2 \times 2 \times 3}{2 \times 2 \times 3 \times 3}$$

$$\frac{12}{36} = \frac{2 \times 2 \times 3}{2 \times 2 \times 3 \times 3}$$

$$\frac{12}{36} = \frac{2 \times 2 \times 3}{2 \times 2 \times 3 \times 3}$$

$$\frac{12}{36} = \frac{2 \times 2 \times 3}{2 \times 2 \times 3 \times 3} = \frac{1}{3}$$

Figure 5-2: Cancelling down fractions.

Cancelling up

Sometimes – particularly when you're comparing, adding or taking away fractions – you want to turn a fraction into one with a *larger* bottom.

You, there, at the back. Stop sniggering!

To turn a fraction into one with a bigger bottom, follow these steps:

1. **Figure out what you need to multiply the bottom you have *by* to get the bottom you want.** You may need to

cancel *down* the fraction you have before you do this step if you can't see anything that works.

2. **Multiply the top and the bottom of your fraction by your answer from Step 1.** You're done.

For instance, if you have a fraction such as $\frac{2}{5}$ and you want to turn the bottom into 30, you multiply top and bottom by 6 to get $\frac{12}{30}$.

Attempting some equivalent fractions questions

1. Cancel down these fractions as far as you can:

(a) $\frac{3}{4}$ (b) $\frac{9}{27}$ (c) $\frac{10}{40}$ (d) $\frac{75}{100}$ (e) $\frac{45}{50}$ (f) $\frac{36}{48}$ (g) $\frac{8}{14}$ (h) $\frac{40}{100}$

2. For each of the following fractions, either cancel them down or say whether they're already cancelled:

(a) $\frac{8}{12}$ (b) $\frac{19}{40}$ (c) $\frac{9}{12}$ (d) $\frac{20}{50}$ (e) $\frac{37}{50}$ (f) $\frac{13}{20}$ (g) $\frac{8}{10}$ (h) $\frac{3}{4}$

3. Cancel each of the following fractions up to have the given bottom. For example, the answer to $\frac{3}{5}$ (10) would be $\frac{6}{10}$.

(a) $\frac{1}{2}$ (6)   (b) $\frac{3}{4}$ (20)   (c) $\frac{4}{5}$ (20)   (d) $\frac{1}{10}$ (50)   (e) $\frac{2}{5}$ (100)
(f) $\frac{1}{7}$ (42)   (g) $\frac{3}{5}$ (15)   (h) $\frac{11}{25}$ (100)

Finding a Fraction of a Number

Finding a fraction of a number is surprisingly easy. Here's what you do:

1. **Divide the number by the bottom of the fraction.**

2. **Multiply your answer from Step 1 by the top of the fraction.** Job done.

So, to find $\frac{2}{3}$ of 36, you divide the number by 3 ($36 \div 3 = 12$) and multiply by 2 ($12 \times 2 = 24$); 24 is two-thirds of 36.

Working backwards

If you're given a number and a fraction and have to work out what your number is the given fraction of – something like '24 is three-quarters of what number?', you have to do exactly the opposite of the previous recipe:

1. **Divide the number by the *top* of the fraction.** Here, you get 8.

2. **Multiply the number by the *bottom* of the fraction.** In this case, you get 32; 24 is three-quarters of 32.

What you did in this recipe is really *dividing a number by a fraction*. In the recipe before, you were *multiplying a number by a fraction*. If you want to know more about multiplying and dividing fractions, see the sidebar later in this chapter. If not, you now know all of the fraction tricks you need for basic maths.

Increasing and decreasing

How about if you're asked to increase or decrease a number by a certain fraction? Doing so is easy, too:

1. **Work out the fraction of the number you're given.**

2. **To work out an increase, add the answer from Step 1 on to the original number.**

3. **To work out a decrease, take the answer from Step 1 away from the original number.**

The processes are really that simple! If you're asked to find out what fraction something has increased or decreased by, follow these steps:

1. **Find the difference between the two numbers – this is the *change*.**

2. **Write down the change, underline it and write the *original* number underneath to make a fraction.**

3. **Cancel this down as far as you can – this is your answer.**

Attempting some fraction of a number questions

1. Work out:

(a) ¾ of 20   (b) ⁹⁄₁₀ of 30   (c) ¼ of 16   (d) ⁵⁄₇ of 14

(e) ³⁄₁₀₀ of 500   (f) ⁷⁄₁₀ of 80   (g) ½ of 48

(h) Three-eighths of thirty-two

2. (a) Twenty-one is three-quarters of what number?

(b) Thirty is half of what number?

(c) A stadium is three-quarters full and contains 6,000 people. How many people can it hold when it's full?

3. What do you get if you:

(a) Increase 20 by a quarter?   (b) Decrease 25 by a fifth?

(c) Increase 90 by a tenth?   (d) Decrease 30 by a tenth?

(e) Increase 21 by a seventh?   (f) Decrease 15 by a third?

(g) Decrease 24 by three-quarters?

(h) Increase 12 by a half?

4. (a) A class of 27 students increases in size to 30. By what fraction did the size of the class increase?

(b) A class of 30 students loses three students. By what fraction did the size of the class decrease?

(c) The petrol tank in your car holds 45 litres. You fill up at the start of a journey. At the end of the journey, you have 30 litres in the tank. What fraction of the tank did your journey use?

Comparing Fractions

Comparing fractions simply means 'putting them into order of size'. You may encounter three levels of difficulty in this kind of question; these are:

- Comparing two fractions with the same bottom (easy)
- Comparing two fractions with different bottoms (harder)
- Comparing several fractions with different bottoms (harder still)

In this section, I run you through how to deal with all three types of question, so that they all end up being easy!

Same or different bottoms

Comparing two fractions with the same bottom is so straightforward it's not even worth a list: you just compare the tops. You can see straight away that $\frac{3}{5}$ is bigger than $\frac{2}{5}$, because the bottoms are the same and three is bigger than two.

When the bottoms are different, though, you can't work out the order quite so easily – for instance, $\frac{3}{4}$ is bigger than $\frac{4}{9}$. This one, you do need a recipe for so follow these steps:

1. **Draw a circle around the bottom of the second fraction and a square around the bottom of the first fraction.**
2. **Multiply the top and bottom of the first fraction by the circled number and write it as a new fraction.** In this case, $\frac{3}{4}$ becomes $\frac{27}{36}$. (Equivalent fractions are at work here – see the earlier section 'Keeping Things Fair': $\frac{3}{4}$ is the same as $\frac{27}{36}$.)
3. **Multiply the top and bottom of the second fraction by the squared number and write it as a new fraction.** Here, $\frac{4}{9}$ becomes $\frac{16}{36}$.
4. **Now the bottoms of the two fractions are the same, and the order is easy to work out. Compare the top of the fraction in Step 2 with the top of the fraction in Step 3.** You can see that $\frac{27}{36}$ is bigger than $\frac{16}{36}$, so $\frac{3}{4}$ is bigger than $\frac{4}{9}$.

Several fractions

When you have several fractions to deal with at once, you need to be careful not to become overwhelmed! Keeping your work neat really pays off here.

My favourite way of comparing several fractions at once – for instance, if you have to put $\frac{3}{7}$, $\frac{1}{3}$ and $\frac{4}{9}$ into size order – is to work them out one pair at a time, following this recipe:

1. **Compare the first fraction with the second fraction, and put a tick above the bigger one.** Here, $\frac{3}{7} = \frac{9}{21}$, which is bigger than $\frac{1}{3} = \frac{7}{21}$, so you put a tick above $\frac{3}{7}$.

2. **Compare the first fraction with each of the fractions to its right, always ticking the winner.** Here, there's only one more to do: $\frac{3}{7}$ against $\frac{4}{9}$, which is $\frac{27}{63}$ against $\frac{28}{63}$, so you tick $\frac{4}{9}$.

3. **Now do the same thing, comparing the second fraction against all the fractions to its right.** Here, there's only one ($\frac{1}{3}$ against $\frac{4}{9}$), and you tick $\frac{4}{9}$.

4. **If the list of fractions is longer, compare the third fraction against the fractions to its right, then the fourth fraction and so on.** Each time, tick the winner.

5. **When you're finished, count up the number of ticks above each fraction.** If you've followed the process correctly, you should have a different number for each.

6. **Arrange the fractions according to the number of ticks next to them.** So, the biggest fraction will have the most ticks, the smallest fraction the fewest and all the others will be in the right order.

Attempting some comparing fractions questions

1. Which is larger:

(a) $\frac{1}{2}$ or $\frac{3}{5}$?   (b) $\frac{3}{10}$ or $\frac{1}{3}$?   (c) $\frac{19}{20}$ or $\frac{9}{10}$?

(d) $\frac{1}{4}$ or $\frac{3}{10}$?   (e) $\frac{1}{2}$ or $\frac{2}{5}$?   (f) $\frac{2}{5}$ or $\frac{3}{8}$?

2. Which is smaller:

(a) $\frac{9}{10}$ or $\frac{8}{9}$? (b) $\frac{2}{5}$ or $\frac{3}{10}$? (c) $\frac{7}{8}$ or $\frac{4}{5}$?

(d) $\frac{1}{8}$ or $\frac{1}{10}$? (e) $\frac{3}{8}$ or $\frac{1}{3}$? (f) $\frac{1}{2}$ or $\frac{12}{25}$?

3. Put these lists of fractions in *ascending* order:

(a) $\frac{1}{2}$, $\frac{1}{3}$, $\frac{1}{4}$, $\frac{1}{5}$ (b) $\frac{2}{3}$, $\frac{3}{4}$, $\frac{5}{6}$, $\frac{7}{8}$

(c) $\frac{1}{2}$, $\frac{2}{5}$, $\frac{4}{7}$, $\frac{3}{8}$ (d) $\frac{1}{9}$, $\frac{3}{20}$, $\frac{1}{10}$, $\frac{2}{15}$

Ascending means climbing: you start with the smallest.

Adding and taking away fractions

Adding and taking away fractions is where maths starts to get hard for quite a lot of people. However, you only need a simple, three-step process:

- Make the bottoms the same
- Add or take away the tops (as the sum asks)
- Cancel the answer down if you can

That's not as hard as you remember it, now is it? In this section, I take you through a few examples.

Adding fractions

If you need to add two fractions together *with the same bottom*, you simply add the tops and leave the bottom alone. If you think about that statement for a moment, it makes sense: if you take three objects and add another two objects (for example), you end up with five objects. Replace 'object' with 'twelfth' (or any other fraction) and the same thing is true: three-twelfths plus two-twelfths is five-twelfths.

That gives you a bit of a clue about how to approach fraction questions when the bottoms aren't the same: you start by making them the same, using the method from the 'Cancelling up' section earlier in this chapter.

A typical fraction question might ask you to work out ½ + ⅓. Beware of this question! The examiner is trying to trick you. He wants you to think, 'Oh! That's obvious!' and to write down ⅖, which is the wrong answer. If you write down ⅖, the examiner will give a hollow laugh, pull out his reddest pen and put a great big cross next to your answer. Don't give him the satisfaction.

Instead, follow these instructions very carefully (and follow along with Figure 5-3):

1. **Write out the fractions.** Put a square around the bottom of the first one and a circle around the bottom of the second one.
2. **Multiply the top and bottom of the first fraction (½) by the number with a circle around it (3).** So, ½ becomes 3/6.
3. **Multiply the top and bottom of the second fraction (⅓) by the number with a square around it (2).** So, ⅓ becomes 2/6. You've made the bottoms the same.
4. **Write down the bottom number with a line above it.**
5. **Add up the tops from Steps 2 and 3 (which happen to be 3 and 2, making 5) and write them above the line.**
6. **Cancel down if you can.** (In this example, you can't.)

$$\frac{1}{\boxed{2}} + \frac{1}{\textcircled{3}}$$

$$\frac{1 \times \textcircled{3}}{2 \times \textcircled{3}} + \frac{1 \times \boxed{2}}{3 \times \boxed{2}}$$

$$\frac{3}{6} + \frac{2}{6} = \frac{5}{6}$$

Figure 5-3: Following the steps to add fractions.

If you're not sure about fractions, drawing cakes to check your answer looks about right is a good idea – especially with the simpler ones. If you draw ½ and ⅓, you can see quite clearly that they don't add up to ⅖ (which is actually less than ½).

Taking away fractions

Taking away fractions works in just the same way as adding fractions, except that you take away the numbers on top instead of adding them.

If the bottoms of the two fractions are the same, you simply take away the tops and leave the bottoms alone (so ⁵⁄₇ – ²⁄₇ = ³⁄₇). If the bottoms are different, such as ½ – ⅙, just follow this recipe:

1. **Write out the fractions.** Put a square around the bottom of the first one and a circle around the bottom of the second one.

Multiplying and dividing fractions

You might remember having to multiply and divide fractions by each other – but the numeracy curriculum doesn't want you to know anything about that. How to multiply and divide fractions aren't secrets or anything; the processes just aren't particularly useful in everyday life. But just in case some smart Alec asks you to multiply or divide two fractions, here's how you do it. To work out ⅔ × ⅘, follow these steps:

1. **Multiply the two top numbers together (2 × 4 = 8).**

2. **Multiply the two bottom numbers together (3 × 5 = 15).**

3. **Write the top number, draw a line underneath it and then write the bottom number (8/15).**

4. **Cancel down if you can.**

Multiplying fractions is easier than adding them, if you ask me. As for dividing, the process is very similar with just one change. To work out ⅔ ÷ ⅘, you need to:

1. **Leave the first fraction alone and flip the *second* fraction upside-down (it becomes ⁵⁄₄).**

2. **Multiply the two fractions using the method above (⅔ × ⁵⁄₄ = ¹⁰⁄₁₂ = ⅚). That's it!**

2. **Multiply the top and bottom of the first fraction ($\frac{1}{2}$) by the number with a circle around it (6).** So, $\frac{1}{2}$ becomes $\frac{6}{12}$.

3. **Multiply the top and bottom of the second fraction ($\frac{1}{6}$) by the number with a square around it (2).** So, $\frac{1}{6}$ becomes $\frac{2}{12}$. You've made the bottoms the same.

4. **Write down the bottom number with a line above it.**

5. **Take away the tops from Steps 2 and 3 (6 − 2 = 4) and write them above the line.**

6. **Cancel down if you can.** You have $\frac{4}{12}$, both of which you can divide by 4 to make $\frac{1}{3}$.

Attempting some adding and taking away fractions questions

1. Work out and cancel down where possible:

(a) $\frac{1}{5} + \frac{2}{5}$ (b) $\frac{1}{12} + \frac{5}{12}$ (c) $\frac{2}{9} + \frac{4}{9}$ (d) $\frac{1}{7} + \frac{5}{7}$ (e) $\frac{4}{5} - \frac{2}{5}$

(f) $\frac{5}{7} - \frac{2}{7}$ (g) $\frac{4}{9} - \frac{1}{9}$ (h) $\frac{9}{10} - \frac{3}{10}$

2. Work out and cancel down where possible:

(a) $\frac{4}{9} + \frac{1}{3}$ (b) $\frac{1}{6} + \frac{2}{3}$ (c) $\frac{1}{30} + \frac{3}{20}$ (d) $\frac{2}{5} + \frac{1}{6}$ (e) $\frac{9}{10} - \frac{2}{5}$

(f) $\frac{2}{5} - \frac{1}{3}$ (g) $\frac{3}{4} - \frac{3}{5}$ (h) $\frac{17}{20} - \frac{4}{5}$

3. Work out:

(a) $\frac{1}{2} \times \frac{4}{9}$ (b) $\frac{19}{20} \times \frac{1}{5}$ (c) $\frac{5}{9} \times \frac{2}{3}$ (d) $\frac{1}{3} \div \frac{3}{4}$ (e) $\frac{3}{20} \div \frac{1}{4}$

(f) $\frac{7}{9} \div \frac{3}{10}$

Working through Review Questions

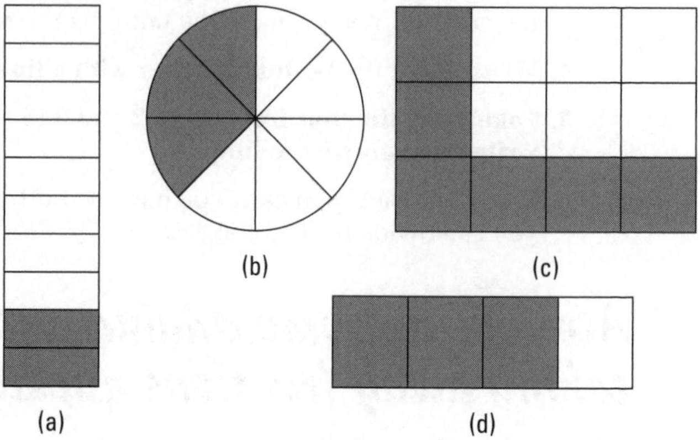

1. Look at the figure above; what fractions do pictures (a) to (d) represent?

2. Draw pictures to represent:

(a) $\frac{2}{5}$ (b) $\frac{3}{8}$ (c) $\frac{1}{4}$ (d) $\frac{1}{2}$ (e) $\frac{7}{10}$ (f) $\frac{1}{3}$ (g) $\frac{3}{4}$ (h) $\frac{2}{3}$

3. Cancel down the following fractions:

(a) $\frac{90}{100}$ (b) $\frac{12}{20}$ (c) $\frac{5}{10}$ (d) $\frac{8}{12}$ (e) $\frac{10}{15}$ (f) $\frac{2}{6}$ (g) $\frac{9}{12}$ (h) $\frac{8}{16}$

4. Work out:

(a) $\frac{3}{4}$ of 100 (b) $\frac{1}{2}$ of 70 (c) $\frac{3}{5}$ of 20 (d) $\frac{1}{4}$ of 40
(e) $\frac{3}{10}$ of 80 (f) $\frac{4}{5}$ of 60 (g) $\frac{9}{10}$ of 30 (h) $\frac{7}{10}$ of 70

5. Calculate (and cancel if possible):

(a) $\frac{7}{10} - \frac{1}{3}$ (b) $\frac{5}{9} + \frac{2}{5}$ (c) $1 - \frac{3}{10}$ (d) $\frac{4}{9} + \frac{1}{10}$
(e) $\frac{5}{6} - \frac{1}{3}$ (f) $\frac{1}{9} + \frac{3}{10}$ (g) $\frac{3}{4} - \frac{7}{10}$ (h) $\frac{1}{4} + \frac{3}{10}$

Checking Your Answers

If you've made it through all of those questions, congratulations! For my money, this is the hardest chapter in the whole book, so well done for getting this far.

Now it's time to check your work. Don't worry if you get some of the answers wrong – it's part of the learning process. Have another look, see if you can spot the mistake – then try again!

Recognising fractions questions

1. (a) ⅖   (b) ³⁄₁₀   (c) ½   (d) ⁹⁄₁₆
   (e) ¾   (f) ⅛

2. (a) ⁴⁄₇   (b) ¾   (c) ⅖   (d) ³⁄₆ (or ½)

(a)

(b)

(c)

(d)

(e)

(f)

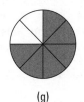
(g)

(h)

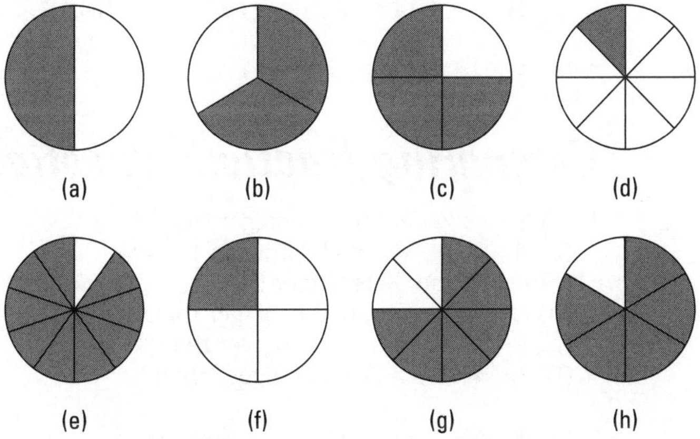

3. See the figure above. If you drew the pictures as rectangles rather than circles, that's fine too.

4. (a) 7   (b) 1   (c) 8   (d) 5   (e) 2   (f) 6   (g) 4   (h) 3

Equivalent fractions questions

1. (a) ½   (b) ⅓   (c) ¼   (d) ¾
   (e) ⁹⁄₁₀   (f) ¾   (g) ⁴⁄₇   (h) ⅖

2. (a) $\frac{2}{3}$ (b) already cancelled (c) $\frac{3}{4}$
   (d) $\frac{2}{5}$ (e) already cancelled (f) already cancelled
   (g) $\frac{4}{5}$ (h) already cancelled

3. (a) $\frac{3}{6}$ (b) $\frac{15}{20}$ (c) $\frac{16}{20}$ (d) $\frac{5}{50}$
   (e) $\frac{40}{100}$ (f) $\frac{6}{42}$ (g) $\frac{9}{15}$ (h) $\frac{44}{100}$

Note: (g) is tricky – you have to cancel it down to $\frac{3}{5}$ before you cancel up to $\frac{9}{15}$.

Fraction of a number questions

1. (a) 15 (b) 27 (c) 4 (d) 10
   (e) 15 (f) 56 (g) 24 (h) 12

2. (a) 28 (b) 60 (c) 8,000

3. (a) 25 (b) 20 (c) 99 (d) 27
   (e) 24 (f) 10 (g) 6 (h) 18

4. (a) $\frac{1}{9}$ (b) $\frac{1}{10}$ (c) $\frac{1}{3}$

Comparing fractions questions

1. (a) $\frac{3}{5} = \frac{6}{10}$; $\frac{1}{2} = \frac{5}{10}$, so $\frac{3}{5}$ is bigger than $\frac{1}{2}$
   (b) $\frac{1}{3} = \frac{10}{30}$; $\frac{3}{10} = \frac{9}{30}$, so $\frac{1}{3}$ is bigger than $\frac{3}{10}$
   (c) $\frac{9}{10} = \frac{18}{20}$, so $\frac{19}{20}$ is bigger than $\frac{9}{10}$
   (d) $\frac{1}{4} = \frac{5}{20}$; $\frac{3}{10} = \frac{6}{20}$, so $\frac{3}{10}$ is bigger than $\frac{1}{4}$
   (e) $\frac{1}{2} = \frac{5}{10}$; $\frac{2}{5} = \frac{4}{10}$, so $\frac{1}{2}$ is bigger than $\frac{2}{5}$
   (f) $\frac{2}{5} = \frac{16}{40}$; $\frac{3}{8} = \frac{15}{40}$, so $\frac{2}{5}$ is bigger than $\frac{3}{8}$

2. (a) $\frac{9}{10} = \frac{81}{90}$; $\frac{8}{9} = \frac{80}{90}$, so $\frac{8}{9}$ is smaller than $\frac{9}{10}$
   (b) $\frac{2}{5} = \frac{4}{10}$, so $\frac{3}{10}$ is smaller than $\frac{2}{5}$
   (c) $\frac{7}{8} = \frac{35}{40}$; $\frac{4}{5} = \frac{32}{40}$, so $\frac{4}{5}$ is smaller than $\frac{7}{8}$
   (d) $\frac{1}{8} = \frac{10}{80}$; $\frac{1}{10} = \frac{8}{80}$, so $\frac{1}{10}$ is smaller than $\frac{1}{8}$
   (e) $\frac{3}{8} = \frac{9}{24}$; $\frac{1}{3} = \frac{8}{24}$, so $\frac{1}{3}$ is smaller than $\frac{3}{8}$
   (f) $\frac{1}{2} = \frac{25}{50}$ $\frac{12}{25} = \frac{24}{50}$, so $\frac{12}{25}$ is smaller than $\frac{1}{2}$.

3. (a) $\frac{1}{5}$, $\frac{1}{4}$, $\frac{1}{3}$, $\frac{1}{2}$ (b) $\frac{2}{3}$, $\frac{3}{4}$, $\frac{5}{6}$, $\frac{7}{8}$ (c) $\frac{3}{8}$, $\frac{2}{5}$, $\frac{1}{2}$, $\frac{4}{7}$
   (d) $\frac{1}{10}$, $\frac{1}{9}$, $\frac{2}{15}$, $\frac{3}{20}$

Adding and taking away fractions questions

1. (a) $\frac{3}{5}$ (b) $\frac{6}{12} = \frac{1}{2}$ (c) $\frac{6}{9} = \frac{2}{3}$ (d) $\frac{6}{7}$
   (e) $\frac{2}{5}$ (f) $\frac{3}{7}$ (g) $\frac{3}{9} = \frac{1}{3}$ (h) $\frac{6}{10} = \frac{3}{5}$

2. (a) $\frac{4}{9} + \frac{1}{5} = \frac{20}{45} + \frac{9}{45} = \frac{29}{45}$ (b) $\frac{1}{6} + \frac{2}{5} = \frac{5}{30} + \frac{12}{30} = \frac{17}{30}$
   (c) $\frac{1}{30} + \frac{3}{20} = \frac{20}{600} + \frac{90}{600} = \frac{110}{600} = \frac{11}{60}$
   (d) $\frac{4}{5} + \frac{1}{6} = \frac{24}{30} + \frac{5}{30} = \frac{29}{30}$ (e) $\frac{9}{10} - \frac{2}{5} = \frac{45}{50} - \frac{20}{50} = \frac{25}{50} = \frac{1}{2}$
   (f) $\frac{3}{5} - \frac{1}{3} = \frac{9}{15} - \frac{5}{15} = \frac{4}{15}$ (g) $\frac{3}{4} - \frac{3}{5} = \frac{15}{20} - \frac{12}{20} = \frac{3}{20}$
   (h) $\frac{17}{20} - \frac{4}{5} = \frac{85}{100} - \frac{80}{100} = \frac{5}{100} = \frac{1}{20}$

3. (a) $\frac{4}{18} = \frac{2}{9}$ (b) $\frac{19}{100}$ (c) $\frac{10}{81}$ (d) $\frac{4}{9}$
   (e) $\frac{12}{20} = \frac{3}{5}$ (f) $\frac{70}{81}$

Review questions

1. (a) $\frac{2}{10} = \frac{1}{5}$ (b) $\frac{3}{8}$ (c) $\frac{6}{12} = \frac{1}{2}$ (d) $\frac{3}{4}$

2. See the figure below.

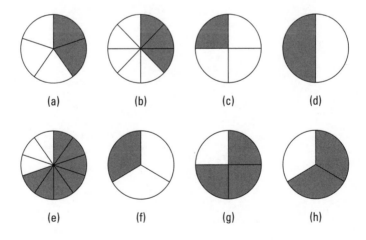

3. (a) $\frac{9}{10}$ (b) $\frac{3}{5}$ (c) $\frac{1}{2}$ (d) $\frac{2}{3}$
   (e) $\frac{2}{3}$ (f) $\frac{1}{3}$ (g) $\frac{3}{4}$ (h) $\frac{1}{2}$

4. (a) 75   (b) 35   (c) 12   (d) 10
(e) 24   (f) 48   (g) 27   (h) 49

5. (a) $\frac{21}{30} - \frac{10}{30} = \frac{11}{30}$ (b) $\frac{25}{45} + \frac{18}{45} = \frac{43}{45}$
(c) $\frac{10}{10} - \frac{3}{10} = \frac{7}{10}$ (d) $\frac{40}{90} + \frac{9}{90} = \frac{49}{90}$
(e) $\frac{15}{18} - \frac{6}{18} = \frac{9}{18} = \frac{1}{2}$ (f) $\frac{10}{90} + \frac{27}{90} = \frac{37}{90}$
(g) $\frac{30}{40} - \frac{28}{40} = \frac{2}{40} = \frac{1}{20}$ (h) $\frac{10}{40} + \frac{12}{40} = \frac{22}{40} = \frac{11}{20}$

Chapter 6

What's the Point? Dealing with Decimals

Decimals have a reputation for being a bit tricky, but I say 'nonsense!' In fact, you probably do more sums with decimals in your day-to-day life than you do with whole numbers.

No, wait, listen, it's true. Most of the sums you do are money-related. And that dot that separates the pounds from the pence? That's a *decimal point*. More precisely, a decimal point is a dot in a number that tells you where the whole numbers end and the parts that are smaller than a whole number start.

That sounds a bit abstract, doesn't it? To make decimals a bit clearer, I first take you through the ideas behind the *decimal system* (which is a fancy name for 'how we write numbers') before I then show you that decimal sums are just like normal sums except that they contain a dot.

Finally, I show you how decimals are related to fractions and how you can change them back and forth to make things easier.

Working with Powers of Ten

Ten is an incredibly useful number in basic maths. In fact, our whole system of writing numbers is based on the number 10 (when you're counting, ten is the first number you need two digits for; one hundred, which is 10×10, is the first number you need three digits for, and so on with one thousand, ten thousand . . . and it's not a coincidence). In fact, whenever you write out a whole number, you can think of the digits, as you go backwards, that is, 'how many ones, how many tens, how many hundreds . . . ?'

In this section, you get to go the other way – beyond the right-hand end of the number. The key thing to remember in this chapter is that there's nothing special about the dot; it's just a marker to show you where the whole number ends and the smaller bits start.

To go beyond the right-hand end of the number, you need to be able to recognise *powers of ten*. Luckily, doing so's really easy. A number is a power of ten in two cases:

- ✔ It's 1, or 1 followed by any number of zeros (like 10, or 1,000, or 1,000,000,000)

- ✔ It's 0.1, or 0. followed by any number of zeros and a 1 (like 0.001 or 0.00000001)

In this chapter, you mainly care about the second case!

Dealing with the dot

I don't go into the details of the place system here – you can check out *Basic Maths For Dummies* (Wiley) if you're that interested – rather, I try to get you to think of numbers in a slightly different way.

Maybe you know that when you multiply a whole number by 10, you just stick a zero on the end – so $45 \times 10 = 450$. I want you to think of this process as is shown in Figure 6-1: instead of 'sticking a zero on the end', you've picked up the number and moved everything one space to the left. Because there's an empty space, you fill it with a zero.

The tricky bit is going the other way: when you pick up the number to move it right, you fall off the end of the number! The way to get around this is to mark the 'end' of the number with a dot – a *decimal point* – to show where the whole number ends and smaller bits begin. In this chapter, I encourage you to turn the dot into a line to help you keep everything lined up.

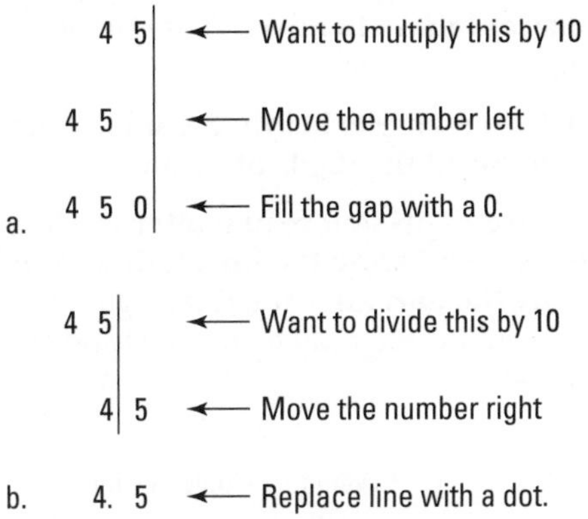

```
       4  5 |   ◄──── Want to multiply this by 10


    4  5    |   ◄──── Move the number left

a.  4  5  0|   ◄──── Fill the gap with a 0.


    4  5|       ◄──── Want to divide this by 10

       4| 5     ◄──── Move the number right

b.  4. 5       ◄──── Replace line with a dot.
```

Figure 6-1: (a) Moving a number to the left; (b) Moving a number to the right.

Multiplying by powers of ten

Multiplying a whole number by ten is one of the easiest times sums you could possibly ask for. When you multiply by ten, you pick the number up and move it one place to the left.

How about multiplying by 100? Well, that's the same as multiplying by ten and then multiplying by ten again – so you could just pick the number up and move it *two* places to the left. A pattern is emerging here, right? Here's how you multiply by any power of ten that's bigger than 1 (for instance, 2.45×1000):

1. **Write down the number that's not a power of ten, and draw a line through the dot and down onto the next line.**

2. **Count how many zeros are in the power of ten.** Here, it's three.

3. **Mentally pick up all the digits in the original number (245) and move them left by the number of spaces in Step 2.** You should now have 245 and a space before the line.

4. **If you have any gaps between your number and the line, fill them with zeros.** You have 2450 before the line.

5. **If you have no digits after the line, just write down the answer before the line.** That number is your answer.

6. **If you have no digits before the line, write down 0. followed by all the digits after the line.**

7. **If you have digits before and after the line, write down the part before the line, followed by a dot, followed by the part after the line.** Figure 6-2 provides examples of (a) multiplying by 1000 and (b) multiplying by 100.

a.

```
2 4.│3 7 5   ◄──── Want to multiply this by 100

2 4 3 7│5   ◄──── Move it two spaces to the left

  2437.5    ◄──── Replace line with a dot
```

b.

```
      2.│4 5   ◄──── Want to multiply this by 1000

  2 4 5 │      ◄──── Move it three spaces left

  2 4 5 0      ◄──── Fill the gap with a zero
```

Figure 6-2: Examples of multiplying by the power of 10: (a) Multiplying by 100; (b) Multiplying by 1000.

Dividing by powers of ten

Dividing by ten (and powers of ten) is pretty much the same thing but the other way around – instead of moving the number left to make it bigger, you move it right to make it smaller. Here's the recipe – and a couple of examples are shown in Figure 6-3 – working through 34.5 ÷ 100:

1. **Write down the number that's not a power of ten, and draw a line through the dot and down onto the next line.**

2. **Count how many zeros are in the power of ten.** Here, it's two.

3. **Mentally pick up all the digits in the original number (345) and move them right by the number of spaces in Step 2.** You should now have 345 straight after the line.

4. **If you have any gaps between your number and the line, fill them with zeros.** Also, if there's nothing before the dot, it's good practice to put a zero there as well.

5. **Write down the number with the dot where your line is.** That number is your answer.

3 4│5   ←—— Want to divide this by 100

│ 3 4 5 ←—— Move it two spaces to the right

   0.345   ←—— Fill the gap with a zero
a.   and replace the line with a dot

1 9 7│4   ←—— Want to divide this by 1000

│ 1 9 7 4 ←—— Move it three spaces right

  0.1974   ←—— Fill the gaps with zeroes
b.   and replace the line with a dot

Figure 6-3: Dividing by powers of 10: (a) Dividing by 100; (b) Dividing by 1000.

Putting a zero in front of the dot is good practice because it's quite easy to miss a dot at the start of a number. If you put the zero in there, it highlights that something funny is going on and helps you to spot the dot.

Attempting some powers of ten questions

1. Work out:

(a) 25×10 (b) 2.5×10 (c) 0.25×10
(d) 0.025×10 (e) $490 \div 10$ (f) $49 \div 10$
(g) $4.9 \div 10$ (h) $0.49 \div 10$

2. Work out:

(a) 354×100 (b) 73.5×100 (c) 5.38×100
(d) 0.054×100 (e) 5.387×100 (f) $94.3 \div 100$
(g) $0.262 \div 100$ (h) $0.6 \div 100$ (i) $9153 \div 100$

3. Fill in the blank:

(a) $649.5 \div \_\_\_ = 6.495$ (b) $1.156 \times \_\_\_ = 11{,}560$
(c) $\_\_\_ \div 10 = 25.48$ (d) $\_\_\_ \times 100 = 188$

Diving into Decimal Sums

As I've said 100.1 times already in this chapter, you only need to know two things about doing sums with decimals:

- ✔ You need to be a bit careful about where the dot is.

- ✔ Other than that, decimal sums are just like normal sums.

To clarify what I mean when I say 'be a bit careful about where the dot is': when you do sums with whole numbers, you need to lay things out so the ends of the numbers – in the 'ones' column – are in line with each other. You do so to make sure that the numbers you add at each stage are the same size, that is, so you're adding hundreds to hundreds and tens to tens.

The process is actually the same with decimals. When you add and take away, you need to make sure that the dots are above each other; doing so ensures that the numbers in each column are the same size (the tens are all in the same column, as are the thousands and, for that matter, so are the tenths and the thousandths after the dot).

Adding and subtracting decimals

In detail, then, here's how you add up two decimals numbers. Let's say we have to work out $160.729 + 22.4$ (in Figure 6-4(a)) and $77 + 19.48$ (Figure 6-4(b)):

1. **Write out the first number.** If it has a dot in it, draw a vertical line through the dot onto the next line, like the one in Figure 6-4(a). If it doesn't have a dot, draw the

line at the end of the number instead, like the one in Figure 6-4(a).

2. **If the second number has a part after the dot, write it beneath the first number, after the line.** Here, the 4 goes after the line on the second row.

3. **Write the whole number part of the second number beneath the first one so it finishes up against the line.** Now you have your numbers lined up correctly!

4. **Add up the numbers as you would for whole numbers, completely ignoring the line for now.**

5. **Put a dot in your answer, on the vertical line, as in Figures 6-4(a) and 6-4(b).** You're done.

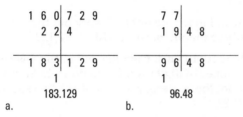

a.   b.

Figure 6-4: Adding decimal numbers: (a) When both numbers have a decimal point; (b) When only one number has a decimal point.

Multiplying and dividing decimals

In basic maths, you don't have to do anything super-fancy like multiplying decimals by other decimals – you have that to look forward to if you go on to study maths further (and I hope you will). Instead, all you need to be able to do is multiply and divide decimals by whole numbers.

That situation is actually a stroke of luck because the process is no harder than multiplying or dividing normal numbers. Here's the recipe for working out 1.234×7:

1. **Count the number of digits your decimal number has after the dot, and write it down in a circle.** Here, it's 3.

2. **Now, ignore the dot and work out the sum using whole numbers.** So, $1,234 \times 7 = 8,638$.

3. **Now count back from the end of the answer the number of digits you wrote down in Step 1, and put a dot there.** Here, you count back three spaces and get 8.638.

4. **Check your answer makes sense, by rounding off each of the numbers and doing the sum roughly.** Here, $1 \times 7 = 7$, so the answer is in the right ballpark – it's a little way off, but it's not 80 or 0.8, which would be obviously wrong.

Dividing is almost exactly the same. Let's work out $169.76 \div 8$:

1. **Count the number of digits your decimal number has after the dot, and write it down in a circle.** Here, it's 2.

2. **Now, ignore the dot and work out the sum using whole numbers.** So, $16,976 \div 8 = 2,122$.

3. **Now count back from the end of the answer the number of digits you wrote down in Step 1, and put a dot there.** Here, you count back two spaces and get 21.22.

4. **Check your answer makes sense, by rounding off each of the numbers and doing the sum roughly.** Here, I might try $170 \div 10 = 17$ or $160 \div 8 = 20$, both of which are easy to work out and aren't far off the final answer.

Going off the end

If you're dividing and you get to the end of your number and still have a remainder, you have a problem. Luckily, it's a problem that can usually be solved using decimals. The key thing to remember is that you can add as many zeros as you like after a decimal point without changing the number: 16 is the same as 16.0000 and 15.4 is the same as 15.40 and 15.400000.

So, the solution to dividing a number when 'it goes off the end' is just to add zeroes to the number you're dividing until you can get through the sum without a remainder. (Adding three zeros is almost always enough in basic maths.)

Figure 6-5 demonstrates a couple of examples of this process.

$$\begin{array}{r} 8\,|\,2\ 5 \\ 2\overline{)\,1\ ^16\,|\,5\ ^10} \\ 8.25 \end{array}$$

a.

$$\begin{array}{r} 2\ 4\,|\,1\ 2\ 5 \\ 8\overline{)\,1\ ^19\ ^33\,|\,^10\ ^20\ ^40} \\ 24.125 \end{array}$$

b.

Figure 6-5: Dividing 'off the end' of decimals: (a) 16.5 ÷ 2; (b) 193 ÷ 8.

Attempting some decimal sum questions

1. Work out:

(a) 22.64 + 47.27   (b) 29.219 + 41.29   (c) 102.1 + 15.22
(d) 81 + 30.82   (e) 30.9 + 0.663   (f) 99.9 + 0.109
(g) 50.654 + 6.63   (h) 16.5 + 43.57

Working out a rough estimate for your answer either before or after doing a sum is a good idea so that you know you're getting something close to the right answer.

2. Work out:

(a) 88.287 − 77.854   (b) 84.84 − 51.4   (c) 40.05 − 36.133
(d) 90.3 − 11.928   (e) 100 − 0.01   (f) 75.16 − 4.806
(g) 95.72 − 6.1378   (h) 53 − 0.2234

3. Work out:

(a) 2.539 × 4   (b) 96.07 × 2   (c) 50.803 × 8
(d) 0.145 × 7   (e) 0.999 × 5   (f) 0.001 × 1500
(g) 62.6 × 12   (h) 60 × 1.12

4. Work out:

(a) 82.98 ÷ 2   (b) 90.75 ÷ 3   (c) 29.844 ÷ 4   (d) 4.815 ÷ 5
(e) 33.636 ÷ 6   (f) 98.9898 ÷ 7   (g) 0.8784 ÷ 8   (h) 1.827 ÷ 9

5. Work out:

(a) 9.3 ÷ 1000   (b) 9.3 × 1000   (c) 745.3 × 100
(d) 745.3 ÷ 10,000.   (e) 32.42 × 10,000   (f) 32.42 ÷ 10,000

Converting Decimals

I have a confession to make. I've been keeping a secret from you. I hope you'll forgive me. In this section, I reveal all.

The big secret is that decimals and fractions are really the same thing. They're just different ways of expressing bits of a number that are smaller than one.

In this section, I show you how to convert back and forth between fractions and decimals even if you don't have a calculator.

Fixing fractions

Converting a fraction into a decimal is straightforward. Actually, I sneakily had you practise doing so earlier in the chapter (hopefully) without you noticing. Here's the recipe; I use it to work out ⁷⁄₂₀ as a decimal:

1. **Write down the top number, followed by a dot and as many zeroes as you think you'll need.** For basic maths, three is normally enough, but you can add more if you need to. I write down 7.000.

2. **Divide your number by the bottom.** That's all there is to it: ⁷⁄₂₀ = 7.000 ÷ 20 = 0.35.

A fraction is simply a division sum in disguise!

Dealing with a recurring problem

One problem exists with this method, though, and it comes up when the bottom of the fraction is . . . not very nice. Not very nice numbers, such as 3 and 7, lead to a headache because no matter how many zeroes you add, you always end up with a remainder. (As an exercise, try converting ⅓ to a decimal – but don't go beyond a hundred or so zeros; hopefully you'll see a pattern before then!)

This subject is straying a little beyond basic maths, but I want you to be aware of it: when the bottom of a cancelled-down fraction is a not very nice number, you end up with a *recurring decimal.* When you divide 1 by 3, you get 0.333 . . . going on forever. To demonstrate that this is a repeating decimal, you can write 0.3̇, with a dot over the 3 to show that it repeats. If you worked out ⁵⁄₁₁, you'd get 0.454545. . ., which you can write as 0.4̇5̇, with a dot over both of the repeating numbers. You can even get things like ⅙ = 0.1666 . . . , which you can write as 0.16̇. Decimals that aren't recurring are called *terminating decimals.*

You're unlikely to come across anything harder than recurring decimals in a test, but you may see them on your calculator or somewhere else and wonder what the dot is all about.

You can do a test to see if a number is 'very nice' or not. If it's even, you divide it by two, and keep dividing by two until it's an odd number. Once you have an odd number, if it ends in five, divide it by five until it doesn't end in five any more. If you end up with one, it's a 'very nice' number that you can write as a terminating decimal; if it's anything else, it's a 'not very nice' number that gives you a recurring decimal.

Converting decimals to fractions

Converting decimals to fractions isn't quite as easy, but it does have a simple recipe. Here's what you do to work out what 1.875 is as a fraction:

1. **If there's a number before the dot, just ignore it for the moment.**

2. **Write down the number after the dot as the top of a fraction.** For this one, it's 875.

3. **Count how many digits are in the number in Step 2, and write down 1 followed by that many zeroes as the bottom of the fraction.** Here, there are three digits, so the bottom of the fraction is 1000.

4. **Cancel down the fraction.** So, ⁸⁷⁵⁄₁₀₀₀ = ¹⁷⁵⁄₂₀₀ = ³⁵⁄₄₀ = ⅞.

5. **Write down the number you ignored in Step 1 followed by the answer from Step 4.** So, 1.875 is the same as 1 ⅞.

Attempting some converting decimals questions

1. Write the following as fractions in their simplest form:

(a) 0.01   (b) 0.3   (c) 0.75   (d) 0.375
(e) 0.95   (f) 0.32   (g) 0.99   (h) 0.001

2. Write the following as decimals:

(a) $\frac{1}{10}$ (b) $\frac{1}{4}$ (c) $\frac{1}{8}$ (d) $\frac{1}{40}$
(e) $\frac{3}{4}$ (f) $\frac{7}{10}$ (g) $\frac{5}{8}$ (h) $\frac{13}{20}$

3. Which of the following fractions can be written as terminating decimals?

(a) $\frac{1}{5}$ (b) $\frac{5}{6}$ (c) $\frac{2}{7}$ (d) $\frac{3}{8}$ (e) $\frac{5}{18}$

4. Match up the following recurring decimals with their fractions:

(a) $\frac{1}{3}$ (b) $\frac{1}{9}$ (c) $\frac{1}{7}$ (d) $\frac{8}{9}$.

(A) 0.142857. . . (B) 0.111. . . (C) 0.888. . . (D) 0.333. . .

5. What are the following as decimals?

(a) 4 $\frac{1}{4}$ (b) 15 $\frac{1}{10}$ (c) 9 $\frac{1}{2}$ (d) 103 $\frac{3}{4}$

Working through Review Questions

1. Work out:

(a) 15×10 (b) 15×100 (c) $15 \div 10$ (d) $15 \div 100$
(e) 0.15×10 (f) 0.15×100 (g) $0.15 \div 10$ (h) $0.15 \div 100$

2. Work out:

(a) 2.35×100 (b) $45.6 \div 10$ (c) 19.99×10 (d) 0.01×1000
(e) 44.404×100 (f) $89 \div 1000$ (g) $89.9 \div 1000$ (h) 9.09×10

3. Work out:

(a) $14.46 + 24.53$ (b) $41.238 + 52.071$ (c) $67.6 + 99.8$
(d) $96.162 + 44.668$ (e) $96.49 - 63.06$ (f) $77.826 - 40.697$
(g) $73.1 - 10.5$ (h) $27.78 - 3.78$

4. Work out:

(a) $86 + 44.9$ (b) $5.9 + 14.35$ (c) $83.764 + 14.34$
(d) $6.6 + 15.6201$ (e) $638.74 - 448.9$ (f) $332.47 - 137.384$
(g) $92 - 62.97$ (h) $47.69 - 15.5$

5. Write the following as fractions in their lowest form:

(a) 0.2 (b) 0.5 (c) 0.34 (d) 0.625
(e) 0.85 (f) 0.75 (g) 0.55 (h) 0.8

6. Write the following as decimals:

(a) $\frac{1}{20}$ (b) $\frac{83}{100}$ (c) $\frac{19}{20}$ (d) $\frac{3}{5}$
(e) $\frac{1}{2}$ (f) $\frac{39}{40}$ (g) $\frac{1}{8}$ (h) $\frac{3}{4}$

7. Match the following recurring decimals with their fractions:

(a) $\frac{2}{3}$ (b) $0.3636\ldots$ (c) $\frac{5}{9}$ (d) $\frac{5}{12}$
(e) $\frac{4}{11}$ (f) $0.41666\ldots$ (g) $0.6666\ldots$ (h) $0.5555\ldots$

8. Put the following decimals into ascending order:

(a) 10.902, 10.092, 100.92, 109.02, 109.2

(b) 0.001, 0.012, 0.101, 0.011, 0.021

(c) 5.321, 5.123, 12.35, 12.53, 15.32

(d) 100.09, 100.1, 100.11, 101.01, 100.9

Checking Your Answers

If your answers are different to mine, a few possible reasons may account for that:

- Maybe I've made a terrible mistake that's been missed by everyone at Dummies Towers. While this is technically possible (I do make a lot of mistakes), it's really unlikely (my editors very rarely do).

- Maybe you haven't lined up the dots properly before doing your sum. If this happens, you normally end up with an answer that's wildly different to the right one – the digits don't even look the same.

- Maybe you've moved a dot the wrong way or the wrong distance. This happens a lot (moving the dot one space when you mean to move it two is easily done). In this case, your answer will look kind of right, but the dot will be in the wrong place.

Powers of ten questions

1. (a) 250   (b) 25   (c) 2.5   (d) 0.25
   (e) 49   (f) 4.9   (g) 0.49   (h) 0.049

2. (a) 35400   (b) 7350   (c) 538   (d) 5.4
   (e) 538.7   (f) 0.943   (g) 0.00262   (h) 0.006
   (i) 91.53

3. (a) 100   (b) 10,000   (c) 254.8   (d) 1.88

Decimal sums questions

1. (a) 69.91   (b) 70.509   (c) 117.32   (d) 111.82
   (e) 31.563   (f) 100.009   (g) 57.284   (h) 60.07

2. (a) 10.433   (b) 33.44   (c) 3.917   (d) 78.372
   (e) 99.99   (f) 70.354   (g) 89.5822   (h) 52.7766

3. (a) 10.156   (b) 192.14   (c) 406.424   (d) 1.015
   (e) 4.995   (f) 1.5   (g) 751.2   (h) 67.2

4. (a) 41.49　(b) 30.25　(c) 7.461　(d) 0.963
   (e) 5.606　(f) 14.1414　(g) 0.1098　(h) 0.203.

5. (a) 0.0093　(b) 9,300　(c) 74,530　(d) 0.07453
   (e) 324,200 (f) 0.003242

Converting decimals questions

1. (a) ¹⁄₁₀₀　(b) ³⁄₁₀　(c) ¾　(d) ⅜
   (e) ¹⁹⁄₂₀　(f) ⁸⁄₂₅　(g) ⁹⁹⁄₁₀₀　(h) ¹⁄₁₀₀₀

2. (a) 0.1　(b) 0.25　(c) 0.125　(d) 0.025
   (e) 0.75　(f) 0.7　(g) 0.625　(h) 0.65

3. (a) Terminating (0.2)　　　(b) Recurring (0.8333...)
   (c) Recurring (0.285714...)　(d) Terminating (0.375)
   (e) Terminating (0.5)

3(c) may not look like a recurring decimal, but those six digits repeat forever. Also, although 18 is a not very nice number, the fraction isn't in its simplest form – it's ½.

4. (a) ⅓ = 0.333... (D)

(b) ⅑ = 0.111... (B)

(c) ⅐ = 0.142857...(A)

(d) ⁸⁄₉ = 0.888... (C)

5. (a) 4.25　(b) 15.1　(c) 9.5　(d) 103.75

Review questions

1. (a) 150　(b) 1500　(c) 1.5　(d) 0.15
   (e) 1.5　(f) 15　(g) 0.015　(h) 0.0015

2. (a) 235　(b) 4.56　(c) 199.9　(d) 10
   (e) 4440.4 (f) 0.089　(g) 0.0899　(h) 90.9

3. (a) 38.99　(b) 93.309　(c) 167.4　(d) 140.83
   (e) 33.43　(f) 37.129　(g) 62.6　(h) 24

4. (a) 130.9   (b) 20.25   (c) 98.104   (d) 21.2201
    (e) 189.84   (f) 195.086   (g) 29.03   (h) 32.19

5. (a) $\frac{1}{5}$ (b) $\frac{1}{2}$ (c) $\frac{17}{50}$ (d) $\frac{5}{8}$
    (e) $\frac{17}{20}$ (f) $\frac{3}{4}$ (g) $\frac{11}{20}$ (h) $\frac{4}{5}$

6. (a) 0.05   (b) 0.83   (c) 0.95   (d) 0.6
    (e) 0.5   (f) 0.975   (g) 0.125   (h) 0.75

7. (a) $\frac{2}{3}$ = (g) 0.666. . .   (b) 0.3636... = (e) $\frac{4}{11}$
    (c) $\frac{5}{9}$ = (h) 0.555. . .   (d) $\frac{5}{12}$ = (f) 0.416. . .

8. (a) 10.092, 10.902, 100.92, 109.02, 109.2

(b) 0.001, 0.011, 0.012, 0.021, 0.101

(c) 5.123, 5.321, 12.35, 12.53, 15.32

(d) 100.09, 100.1, 100.11, 100.9, 101.01

Chapter 7

It's All Relative: Tackling Ratio and Proportion

*Y*ou've seen this scenario in cartoons: the bad guys have returned to their hide-out with a sack of loot and are dividing it between them – one for you, two for me, one for you, two for me . . . whether they know it or not, the villains are using *ratios* to split up their ill-gotten gains. In that example, they're splitting up the money in a ratio of 1:2 – for every one you get, I get two.

When you start looking for ratios, you see them all over the place:

▶ Your bank might say 'for every hundred pounds you invest, you'll get four more at the end of the year' – a ratio of 100:4.

▶ A news story might say, 'there's one GP for every 450 people in the UK' – a ratio of 1:450.

▶ A recipe for squash might say, 'for every 10 millilitres of concentrate, add 100 millilitres of water' – a ratio of 10:100.

And so on. (Does anyone really need a recipe for squash, by the way? Even I'm not that hopeless in the kitchen.)

The basic idea of ratios is to describe how a quantity is split between two or more things – so the bank example split money between your investment and your interest; the news example split people between GPs and non-GPs; and the squash example split the drink between concentrate and water.

Ratios have a lot in common with fractions (see Chapter 5 if fractions make you feel giddy!) but in some ways they're easier. One of the nice things about ratios is that they work really well with a tool called the Table of Joy, which I introduce immediately below. I also show you in this chapter how to cancel down ratios (hint: just like fractions), how to use the Table of Joy to solve ratio questions and how to make life easier with scaling recipes and maps.

The Table of Joy isn't the only way to work out ratio and proportion sums, but it is a consistent method that always works. If you know a better way, go ahead and use it – I won't be hurt. Much. Sniff.

Taking on the Table of Joy

The *Table of Joy* isn't a new idea (or even really my idea). It's based on an old idea called the 'rule of three', but the rule of three doesn't have a cool table to go with it.

It used to be called just 'The Table' until my student, Tain, decided that that was a boring name and came up with the Table of Joy.

The idea of the Table of Joy is to turn any sum dealing with *proportions* into a simple sum that always involves the same steps – so if you can do one kind of Table of Joy sum, you can do them all.

You'll be amazed by how many questions in an average numeracy test you can work out using the Table of Joy. You can use it for:

- ✔ Percentages
- ✔ Ratios

- ✔ Proportion (scaling recipes)
- ✔ Map scales
- ✔ Pie charts
- ✔ Speed–distance–time questions
- ✔ Converting almost anything (except temperatures)

And many more. (If you go on to do a higher-tier GCSE – and I hope you will – you can use it for similar triangles, the sine rule, density . . . the list is almost endless.)

Working out a sum with the Table of Joy involves three stages:

1. **Drawing it out.**
2. **Filling it in and writing the sum.**
3. **Doing the sum.**

All of the stages are short and easy, and I explain them in the next two sections.

Drawing it out and filling it in

Figure 7-1 is a Table of Joy that I constructed to solve the following problem:

Chris and Steve share some money in the ratio 2:5. Chris gets £100; how much does Steve get?

Here's the recipe for drawing it out and filling it in:

1. **Draw a noughts and crosses grid.** Make it big enough to write in comfortably; don't worry about how much space you use up.

2. **In the top row, label what you're measuring.** In this case, it's Chris's money and Steve's money, so I write Chris and Steve or C and S, depending on how much of a hurry I'm in.

3. **In the left column, label the information you have.** Here, the ratio and the actual amount of money.

4. **Fill in the numbers.** In the ratio, the 2 applies to Chris and the 5 to Steve, so write the 2 in the Ratio row under Chris, and the 5 in the same row under Steve. Also, Chris has £100, so that goes in the money row under Chris.

Doing the sum

	Chris	Steve
Ratio	2	5
Money	100	

$$\frac{5 \times 100}{2} = 250$$

Figure 7-1: Using the Table of Joy: (a) Filling it in and shading the squares; (b) Working out the sum.

To find the sum using the Table of Joy, follow these steps:

1. **Shade the squares like a chessboard.** See Figure 7-1(a). You'll have either two numbers on shaded squares or two numbers on unshaded squares.

2. **Write these numbers down with a × sign between them.** For the example in Figure 7-1(a), I write 5 × 100.

3. **Write a ÷ followed by the other number (the one that's on the other-coloured square).** That's the Table of Joy sum! Here, it's 5 × 100 ÷ 2, as shown in Figure 7-1(b).

4. **Work out the sum that's your answer.** I get 500 ÷ 2 = £250.

That process is all there is to using the Table of Joy. You follow the same process every time, whether you're doing percentages, pie charts or proportions.

Attempting some Table of Joy questions

a.

	Chris	Steve
Ratio	2	5
Money		20

b.

	Chris	Steve
Ratio		7
Money	14	49

c.

	Chris	Steve
Ratio	1	
Money	50	200

d.

	Chris	Steve
Ratio	3	4
Money	90	

Figure 7-2: Using the Table of Joy for working out parts.

1. The figure above shows four Table of Joy grids (a, b, c and d) with the numbers filled in. Write down the Table of Joy sum for each grid.

2. Work out the answer to each sum in Question 1.

Rattling Off Ratios

A *ratio* is simply a pair of numbers showing you the relative size of two things – usually shares of something such as money or sweets.

Here's what I mean by 'relative size': if you say 'Dick and Jane split the money in the ratio of 1:2', Jane got twice as much money as Dick.

In this section, I show you how to simplify ratios, and how to do the main types of ratio sum – all of which are easy with the Table of Joy but can also be worked out using other methods. As usual, whichever way works best for you is the right way to do it!

Simplifying ratios

Simplifying a ratio (or putting a ratio in its lowest form, as some exams put it) is a very similar process to cancelling down fractions (a process described in Chapter 5). You simplify a ratio because understanding small numbers is usually easier. Rather than saying 'I got 36 sweets and you got 54', for example, I might say 'we split the sweets in the ratio 2:3', which gives you a better idea of how unfair the deal was for me!

You can spot whether a ratio is in its lowest form by asking, 'is there a number you can divide both parts of the ratio by?' If there is, the ratio isn't in its lowest form! Have a look at the section in Chapter 3 on Cancelling Down if you need to.

Here's how you cancel down a ratio (hint: it's a very similar method to the one for cancelling down fractions described in Chapter 5). Follow these steps:

1. **Find a number you can divide the left and the right number by.** I usually look for 10s first, and then 2s, because they're easy to spot.

2. **Divide the left number and the right number by the number you found in Step 1.**

3. **Repeat Steps 1 and 2 until you can't find anything more to divide by.** What you're left with is your answer.

For the example above (which started as 36:54 sweets), I might spot that I can divide both numbers by 2 to get 18:27, and then both numbers by 3 to get 6:9, and by 3 again to get to the final answer of 2:3.

Working out parts

The commonest type of ratio question (after cancelling down, perhaps) is to tell you a ratio and the size of one thing, and ask you to work out the other. Similarly, you may be given the total and the ratio and have to work out the size of either (or both) parts. In either case, the Table of Joy comes to the rescue. For example, you might have a question that reads: 'Janet and John divide some money in the ratio of 3:4. Janet receives £240; how much does John get?' Here's what you do:

1. **Draw out a big noughts and crosses grid, leaving plenty of room for labels.**

2. **Label the table.** In the two right-most columns, write 'Janet' and 'John'; the middle row is for 'Ratio' and the bottom row is for 'Money'.

3. **Fill in the numbers you know.** Janet/Ratio is 3, John/Ratio is 4 and Janet/Money is 240, as shown in Figure 7-3.

4. **Shade the table like a chessboard.** The 4 and 240 are both on unshaded squares, so the sum is $4 \times 240 \div 3$, as shown in Figure 7-3.

5. **Work out the sum.** So, $4 \times 240 = 960$; $960 \div 3 = 320$.

If you're given a total rather than one of the shares, you follow almost exactly the same process, except that the right-most column is 'Total' and you fill in the total of the two parts of the ratio in the Ratio/Total cell, as you'd expect.

Figure 7-3 shows you how to work out the answer to: 'Steve and Phil have 800ml of curry sauce that they want to split in a ratio of 3:1. How much will Steve get?' The next section covers working with totals.

	Janet	John
Ratio	3	4
Money	240	

$$\frac{4 \times 240}{3} = 320$$

	Steve	Total
Ratio	3	4
Curry		800

$$\frac{3 \times 800}{4} = 600$$

Figure 7-3: Using the Table of Joy to work out totals.

Working out totals

If you're given a ratio and somebody's share and then asked to work out the total number – for instance, if Rob and Nazim share some money in the ratio of 2:3 and Nazim gets £360, how much money was there to begin with? – you crack out the Table of Joy again. So:

1. **Draw out a big noughts and crosses grid, leaving plenty of room for labels.**

2. **Label the table.** In the two right-most columns, write 'Nazim' and 'Total'; the middle row is for 'Ratio' and the bottom row is for 'Money'.

3. **Fill in the numbers you know.** Nazim/Ratio is 3, Total/Ratio is 5 (just the two shares added up – here, 2 + 3) and Nazim/Money is 360.

4. **Shade the table like a chessboard.** The 5 and 360 are both on unshaded squares, so the sum is $5 \times 360 \div 3$.

5. **Work out the sum.** So, $5 \times 360 \div 3 = 1800 \div 3 = 600$.

Working out three-part ratios

The most complicated kind of ratio question you're likely to see in an exam involves a three-part ratio. So, instead of two numbers with a colon between them, you see three, like this: 4:2:1.

All this means is the quantity in question has been split into three unequal parts (rather than two), and for every one thing the last person gets, the middle person gets two and the first person gets four. There's good news, though: everything you know about two-part ratios works just the same for three-part ratios. (You almost never see four or more-part ratios in any context, let alone basic maths, but they would also work in the same way.)

So, for example, if Rod, Jane and Freddie split 200 sweets in the ratio 1:3:6 and you want to know how many sweets Freddie gets, you approach it in just the same way as for a two-person problem; Figure 7-4 shows you the steps. Your Table of Joy would have 'Freddie' and 'Total' in the columns, and 'Ratio' and 'Sweets' in the rows. The Freddie/Ratio cell would contain 6, and Total/Ratio would be 10 (1 + 3 + 6). Total/Sweets would be 200, so the whole sum would be $6 \times 200 \div 10$; if you work that out, you get $1200 \div 10 = 120$.

	Freddie	Total
Ratio	6	10
Sweets		200

$$\frac{6 \times 200}{10} = 120$$

Figure 7-4: Using the Table of Joy to work out a three-part ratio.

Attempting some ratio questions

1. Are the following ratios in their lowest form?

(a) 1:5   (b) 2:3   (c) 4:8   (d) 9:12

(e) 15:30   (f) 9:10   (g) 6:9   (h) 3:5

2. Write the following ratios in their lowest form:

(a) 4:12   (b) 2:12   (c) 3:12   (d) 8:12

(e) 15:20   (f) 18:36   (g) 75:100   (h) 15:45

3. Zippy and Bungle have 360 sweets. How much will Zippy get if they share them in the following ratios:

(a) 1:2   (b) 3:5   (c) 1:9   (d) 2:7

(e) 1:4   (f) 4:5   (g) 2:3   (h) 3:5

4. Find the total number of items if:

(a) They're shared in the ratio of 1:2 and the first person has 20.

(b) The ratio is 1:2 and the second person has 30.

(c) The ratio of 2:5 and the first person has 10.

(d) The ratio is 9:1 and the first person has 45.

5. A cement mixture calls for cement, sand and water in the ratio of 1:2:4.

(a) If there is 10kg of cement, how much sand should there be?

(b) If there is 8kg of water, how much concrete will the mix make?

(c) If you want to make 35kg of concrete, how much sand will you need?

(d) If you used 40kg of sand, how much water would you need?

Scaling Up and Down

Another big favourite of the sadists who set exams is scaling recipes (or anything with ingredients) up and down. For example, if you have a recipe that serves four people but you have eight people coming to dinner, you obviously need to increase the amount of everything you make. In that simple example, you just have to double everything; more likely, though – in an exam, at least – you'll have numbers that don't play nicely together.

This is what maths books mean by 'proportional' – you have to make everything bigger in the same ratios. And when things are proportional, you can use the Table of Joy!

Going bigger and smaller

Let's say you have the ingredients listed below for vegetarian haggis. (I actually made this; it was a disaster.) The recipe is for six . . . but you have 10 guests coming. What quantity of lentils do you need?

Ingredients (serves 6): one onion, one carrot, six mushrooms, 75g lentils, 50g beans, 75g nuts, 50g oats, 250ml vegetable stock.

Follow these steps (to work out the sum, not to make the haggis):

1. **Draw out a noughts and crosses grid, leaving plenty of room for the labels.**

2. **Label the columns 'Servings' and 'Lentils'; label the rows 'Recipe' and 'My meal'.**

3. **Fill in your numbers.** You know that Lentils/Recipe is 75, Servings/My meal is 10 and Servings/Recipe is 6.

4. **Shade in the table like a chessboard and write down the Table of Joy sum.** That is, $75 \times 10 \div 6$.

5. **Work out the sum.** So, $750 \div 6 = 125$ – you need 125g of lentils.

Attempting some scaling questions

1. A recipe for banoffee pie serves four people and suggests the following quantities. If you were cooking for 12, how much of each ingredient would you need?

(a) 135ml condensed milk   (b) 120g digestive biscuits

(c) 50g butter   (d) one banana

2. A recipe for vegetable lasagne serves four people and suggests the following quantities. How much of each ingredient would you need if you were cooking for 10?

(a) 200g lentils   (b) 400g tomatoes

(c) 100g mushrooms   (d) 80g flour

If you need someone to taste-test your banoffee pie, you know where to find me!

3. A parent has a game for four children that requires 4 dice, 12 sheets of paper, 2 pencils, 4 chairs, 10 balloons, 3 litres of lemonade, 8 cakes and 30 counters. She plans to adapt it for six children. Assuming everything increases in proportion to the number of children playing, how many/much will she need of the following:

(a) dice   (b) paper   (c) pencils   (d) chairs,
(e) balloons, (f) lemonade (g) cakes   (h) counters

Magnifying Maps

Working with map scales is no different to working with normal ratios. If you look at a road atlas or an Ordnance Survey map, you'll probably see a ratio on it somewhere saying, for instance, 1:50,000. What that means is: the map has been shrunk by a factor of 50,000 from real life. One centimetre on the map would represent 50,000cm in real life (which works out to be 500m).

Check out Chapter 13 for more about converting between metric measurements.

Only three possible types of map scale questions appear in basic maths:

- ✔ Finding out a distance on the map if you know the scale and the real-life distance.

- ✔ Finding out a distance in real life if you know the scale and the map distance.

- ✔ Finding out the scale if you know a real-life distance and a map distance of the same thing.

In this section, I show you how to work out each of them.

Making it map-sized

To turn a real-life distance into a map distance, follow these steps:

1. **Convert the real-life distance into centimetres.**
   I normally do this by turning kilometres into metres (multiply by 1000) and then metres into centimetres (multiply by another 100). You can use whatever method works for you, though.

2. **If the scale is 1: something, you must divide by the something.** This is your answer in centimetres.

3. **If your scale is some number of kilometres to some number of centimetres (or similar), use the Table of Joy.** Your columns would be kilometres and centimetres, and your rows would be ratio and distance.

For instance, if your scale was '16 miles to the inch', which is bordering on the silly, and you wanted to know how long a 2.5 inch road was in real life, you'd work out the Table of Joy sum to be $2.5 \times 16 \div 1 = 40$ miles.

Making it full-sized

Making a map distance full-sized involves much the same process:

1. **If the scale is 1: something, you must multiply your distance by the something.**

2. **Convert this distance into kilometres.** I normally do this by turning centimetres into metres (divide by 100) and then metres into kilometres (divide by another 1000), but again, use whatever method works for you.

3. **If your scale is some number of kilometres to some number of centimetres (or similar), use the Table of Joy.** Your columns would be kilometres and centimetres, and your rows would be ratio and distance.

Finding the scale

To find the scale, follow these steps:

1. **Make sure the two distances are in the same units by converting (centimetres are usually good).**

2. **Write the two numbers as a ratio.**

3. **If either of the numbers isn't a whole number, multiply both by a number that will make them both whole numbers.**

4. **Cancel the ratio down to its simplest form.**

Attempting some map questions

1. If a map has a scale of 1:50,000, how far on the map is:

(a) A path, if it's 2km in real life?

(b) A stretch of river that's 1500m in real life?

(c) A road that's 7.2km long in real life?

(d) The perimeter of a farm that's 12km long in real life.

2. If a map has a scale of 1:20,000, how far in real life is:

(a) A path that's 2cm long on the map?

(b) A stretch of river that's 15*mm* long on the map?

(c) A road that's 7.2cm long on the map?

(d) The perimeter of a building that's 1.2mm long on the map?

3. What is the scale of each map if:

(a) A path is 3cm long on the map and 6km in real life?

(b) A stretch of river is 8cm long on the map and 4km in real life?

(c) A motorway is 20km long in real life and 20cm on the map?

(d) A ring road that's 20km long in real life and 10cm on the map?

4. Find the lengths of the following things in real life based on these maps:

(a) The scale is 2cm:5km, and a road is 16 millimetres on the map.

(b) The scale is 1 inch:16 miles, and a road is 3 inches on the map.

(c) The scale is 3cm:8km, and a road is 6cm on the map.

(d) The scale is 2cm:15km, and a road is 12 millimetres on the map.

Working through Review Questions

1. Are the following ratios in their lowest form?

(a) 12:18   (b) 6:10   (c) 4:3   (d) 10:9

(e) 13:2   (f) 10:20   (g) 13:16   (h) 4:1

2. Put the following ratios into their simplest form:

(a) 15:20   (b) 12:15   (c) 70:100   (d) 20:18

(e) 15:5   (f) 18:2   (g) 14:16   (h) 3:12

3. Marigold and Peter share a sum of money in the following ratios. In each case, how much does Peter receive?

(a) If the ratio is 1:3 and Marigold receives £150.

(b) If the ratio is 3:1 and Marigold receives £150.

(c) If the ratio is 2:3 and Marigold receives £150.

(d) If the ratio is 3:2 and Marigold receives £150.

(e) If the ratio is 1:2 and the total is £150.

(f) If the ratio is 2:1 and the total is £150.

(g) If the ratio is 2:3 and the total is £150.

(h) If the ratio is 3:2 and the total is £150.

4. Hannah and Arun share some sweets in the following ratios. In each case, how many sweets are there altogether?

(a) If the ratio is 4:1 and Hannah gets 20 sweets.

(b) If the ratio is 1:4 and Hannah gets 20 sweets.

(c) If the ratio is 2:5 and Hannah gets 20 sweets.

(d) If the ratio is 5:2 and Hannah gets 20 sweets.

(e) If the ratio is 2:1 and Arun gets 10 sweets.

(f) If the ratio is 1:2 and Arun gets 10 sweets.

(g) If the ratio is 3:2 and Arun gets 6 sweets.

(h) If the ratio is 2:3 and Arun gets 6 sweets.

5. Kevin, Fergal and Jimmy are sharing out vintage records in the ratio 2:3:4.

(a) If Kevin gets eight records, how many does Jimmy get?

(b) If Fergal gets 12 records, how many records are there altogether?

(c) If there are 90 records, how many does Jimmy get?

(d) If Kevin gets six records, how many are there altogether?

6. Karen is sawing up some wood.

(a) If the pieces are 30cm, 60cm and 120cm long, what is the ratio of the lengths?

(b) If the ratio of lengths is 4:5:6 and the shortest piece is 60cm long, how long is the longest piece?

(c) If the ratio of lengths is 1:3:6, and the original piece of wood is 1 m long, how long is the longest piece?

(d) If the ratio is 2:5:9 and the shortest piece is 10cm long, how long was the original piece of wood before she sawed it up?

7. I have a recipe for 10 people that requires (among other things) 1 kg tomatoes, 5 onions, 150g cheese and 2 litres vegetable stock.

(a) If I cooked the recipe for two people, how much cheese would I need?

(b) If I cooked for 12 people, how many onions would I need?

(c) If I had 600g tomatoes, how many people could I cook for?

(d) If I cooked for four people, how much stock would I need?

8. I have a jelly recipe that requires 200g caster sugar, 600ml water, 500g strawberries and 10g gelatin; it serves four people.

(a) If I have a party of 20 people, how much sugar will I need?

(b) If I'm making 10 portions, how much water will I need?

(c) If I'm just making jelly for myself (that's not weird, right?), what weight of strawberries do I need?

(d) If I have 20g gelatin, how many portions of jelly can I make?

9. If a map has a scale of 1:50,000:

(a) On the map, how long is a road that's 4km in real life?

(b) In real life, how long is a road that's 10cm on the map?

(c) On the map, how long is a stream that's 12km in real life?

(d) In real life, how far is it around a pond that' has a 12cm circumference on the map?

10. If a map has a scale of 1:1,000,000:

(a) On the map, how long is a road that's 10km in real life?

(b) On the map, how long is a motorway that's 200km long?

(c) In real life, how long is a road that's 5.2cm on the map?

(d) In real life, how long is a river that's 1.3cm on the map?

11. What is the scale on each of the following maps?

(a) 10cm on the map represents 25km on the ground.

(b) 1km on the ground is 4cm on the map.

(c) 4cm on the map is 50m on the ground.

(d) 50km on the ground is 8cm on the map.

12. If a map has a scale of 10 miles to the inch:

(a) On the map, how long is a road that's 20 miles in real life?

(b) On the map, how long is a footpath that's 5 miles in real life?

(c) In real life, how long is a motorway that's 5 inches long?

(d) Two stations are 4.5 inches apart on the map. How far apart are they in real life?

Checking Your Answers

If you're having problems with ratio sums, make sure you're setting your work out clearly and labelling everything; don't just write down a number, write down what it is, too. For example, 16 could mean anything, but if you write down 'Map distance: 16cm', it's clear what you mean!

Table of Joy questions

1. (a) $2 \times 20 \div 5$ (b) $7 \times 14 \div 49$ (c) $1 \times 200 \div 50$
(d) $4 \times 90 \div 3$. (You can have the first two numbers either way round, but the last number needs to be after the division sign.)

2. (a) $40 \div 5 = 8$ (b) $98 \div 49 = 2$ (c) $200 \div 50 = 4$
(d) $360 \div 3 = 120$

Ratio questions

1. (a) yes   (b) yes   (c) no (you can divide by 4)
(d) no (you can divide by 3)
(e) no (you can divide by 5)
(f) yes   (g) no (you can divide by 3)   (h) yes

2. (a) 1:3 (divide by 4) (b) 1:6 (divide by 2)
   (c) 1:4 (divide by 3) (d) 2:3 (divide by 4)
   (e) 3:4 (divide by 5) (f) 1:2 (divide by 18; or 2, 3 and 3)
   (g) 3:4 (divide by 25; or 5 and 5)
   (h) 1:3 (divide by 15; or 5 and 3)

3. (a) $1 \times 360 \div 3 = 120$ (b) $3 \times 360 \div 8 = 135$
   (c) $1 \times 360 \div 10 = 36$ (d) $2 \times 360 \div 9 = 80$
   (e) $1 \times 360 \div 5 = 72$ (f) $4 \times 360 \div 9 = 160$
   (g) $2 \times 360 \div 5 = 144$ (h) $3 \times 360 \div 5 = 216$

4. (a) $20 \times 3 \div 1 = 60$ (b) $30 \times 3 \div 2 = 45$
   (c) $10 \times 7 \div 2 = 35$ (d) $45 \times 10 \div 9 = 50$

5. (a) $10 \times 2 \div 1 = 20$kg (b) $8 \times 7 \div 4 = 14$kg
   (c) $35 \times 2 \div 7 = 10$kg (d) $40 \times 4 \div 2 = 80$kg

Scaling questions

1. (a) 405ml (b) 360g (c) 150g (d) 3 bananas

For Question 1, you could either use the Table of Joy or just multiply everything by three – whichever you find easier.

2. (a) 500g (b) 1,000g or 1kg (c) 250g (d) 200g

3. Multiply each of the numbers by 6 and divide by 4 to get:
(a) 6 dice (b) 18 sheets of paper (c) 3 pencils
(d) 6 chairs (e) 15 balloons (f) 4.5 litres of lemonade
(g) 12 cakes (h) 45 counters

Map questions

1. (a) 4cm (b) 3cm (c) 14.4cm (d) 24cm

2. (a) 400m (b) 300m (c) 1440m (d) 24m

3. (a) 1:200,000 (b) 1:50,000 (c) 1:100,000 (d) 1:200,000

4. (a) 4km (b) 48 miles (c) 16km (d) 9km

Review questions

1. (a) No (you can divide both by 2, 3 and 6)
    (b) No (2)   (c) Yes   (d) Yes   (e) Yes
    (f) No (2, 5 and 10)
    (g) Yes   (h) Yes

2. (a) 3:4   (b) 4:5   (c) 7:10   (d) 10:9
    (e) 3:1   (f) 9:1   (g) 7:8   (h) 1:4

3. (a) £450   (b) £50   (c) £225   (d) £100
    (e) £100   (f) £50   (g) £90   (h) £60

4. (a) 25   (b) 100   (c) 70   (d) 28
    (e) 30   (f) 15   (g) 15   (h) 10

5. (a) 16   (b) 36   (c) 40   (d) 27

6. (a) 1:2:4   (b) 90cm   (c) 60cm   (d) 80cm

7. (a) 30g   (b) 6   (c) 6   (d) 800ml

8. (a) 1kg or 1,000g   (b) 1,500ml or 1.5 litres
    (c) 125g (d) 8

9. (a) 8cm   (b) 5km   (c) 24cm   (d) 6km

10. (a) 1cm   (b) 20cm   (c) 52km   (c) 13km

11. (a) 1:25,000   (b) 1:25,000   (c) 1:1,250   (d) 1:625,000

12. (a) 2 inches   (b) ½ inch   (c) 50 miles   (c) 45 miles

Chapter 8

Working Out Perfect Percentages, 100% of the Time

*T*he word 'percentage' comes from the French 'per cent', which means 'for every hundred'. English has many hundred-related words with 'cent' in, such as centimetre (one hundredth of a metre), century (one hundred years, or 100 runs in cricket), centipede (an insect with about 100 legs), centurion (leader of 100 men) and so on.

The main use of percentages is to avoid having to use decimals or fractions! It feels a lot less maths-y to say 'there's a 20% off sale' than to say 'Prices are reduced by a factor of 0.2' or 'prices have been slashed by a fifth'. Using percentages makes the numbers just a little bit easier to think about.

In this chapter, I show you how to convert these easy numbers into (and back from) other forms – such as the decimals and fractions I just mentioned – as well as how to figure out percentages of numbers and, finally, look at some percentage sums you might encounter as you go about your day.

Converting Percentages

Communicating with percentages may be easier, but doing maths – especially with a calculator or computer – is easier using decimals and fractions.

In this section, I show you how to convert back and forth between the different forms. Once you've done a few, you'll get the hang of it.

The nearby sidebar provides a table for finding commonly-used percentages at a glance. If you feel like memorising this table, it will certainly come in useful!

Percentage converter

Here's a handy-dandy table showing percentages that frequently come up and their equivalent decimals and fractions.

Percentage	Fraction	Decimal
1	$\frac{1}{100}$	0.01
2	$\frac{1}{50}$	0.02
5	$\frac{1}{20}$	0.05
10	$\frac{1}{10}$	0.1
20	$\frac{1}{5}$	0.2
25	$\frac{1}{4}$	0.25
40	$\frac{2}{5}$	0.4
50	$\frac{1}{2}$	0.5
60	$\frac{3}{5}$	0.6
75	$\frac{3}{4}$	0.75
80	$\frac{4}{5}$	0.8
90	$\frac{9}{10}$	0.9

Percentages to decimals

Converting a percentage to a decimal is easy. Simply follow this recipe:

1. **If your percentage number is less than 100, add a zero in front of it. If it's less than ten, add another zero at the start.** So, 43 would become 043; 7 would become 007.

2. **If your number doesn't have a dot in it, put one at the end.** So you'd have 043. or 007.

3. **Move the dot two spaces to the left.** Here, you'd have 0.43 or 0.07.

4. **If your number ends with a 0 (after the decimal point), you can ignore it.** If you end up with 0.40, it becomes 0.4.

Decimals to percentages

Converting decimals back to percentages is just as easy. Follow these steps:

1. **If you only have one digit after the dot in your decimal, add another zero at the end.** If you have 0.4, you turn it into 0.40.

2. **Move the decimal point two places to the right.** So, 0.40 becomes 040.

3. **Ignore any zeros at the beginning of the number and write a percentage sign at the end.** This number is your answer – in this case, 40%.

Percentages to fractions

If you know how to cancel down fractions, converting percentages to fractions is a breeze! If you need a refresher on cancelling fractions, have a quick look at Chapter 5.

This recipe only works for whole numbers and *terminating* decimals, ones that don't go on for ever. Luckily, you don't have to care about non-terminating decimals.

Follow these steps:

1. **Write your number with a line underneath it, and the number 100 below that.** So, for example, 17% means exactly the same thing as $^{17}/_{100}$.

2. **If your percentage number isn't a whole number, multiply both the number on top and the number on the bottom by ten; repeat this step until it is a whole number.** So, if you have 12.5/100, it becomes $^{125}/_{1000}$.

3. **Find a number you can divide both the top and bottom by (2 and 5 are good numbers to try). Divide the top and bottom number by this number, and repeat this step until you can't find another number to divide by.** So, $^{125}/_{1000}$ becomes $^{25}/_{200}$, then $^{5}/_{40}$, then $^{1}/_{8}$.

4. **Look at the fraction you're left with.** This is your answer.

Fractions to percentages

Fractions to percentages are probably the most complicated of the topics in this section, but they're still not all that difficult. Here's how they work:

1. **Multiply the top of the fraction by 100, but leave the bottom alone.** For instance, if you have $^{3}/_{20}$, you turn it into $^{300}/_{20}$.

2. **Find a number you can divide both the top and bottom of the fraction by (2, 5 and 10 are good numbers to try if you're stuck). Divide the top and bottom by this number and repeat the process until you can't go any further.** Here, you might divide top and bottom by 10 to get $^{30}/_{2}$, then divide top and bottom by 2 to get $^{15}/_{1}$ = 15.

3. **Put a percentage sign after the number you end up with.** This is your answer. Here, $^{3}/_{20}$ is the same thing as 15%.

Attempting some converting percentages questions

1. Convert these percentages into decimals:

(a) 3%   (b) 70%   (c) 47%   (d) 98%

(e) 53%   (f) 9%   (g) 10%   (h) 15%

(i) 40%   (j) 33%   (k) 42%   (l) 95%

2. Convert these decimals into percentages:

(a) 0.32   (b) 0.88   (c) 0.07   (d) 0.62

(e) 0.6   (f) 0.38   (g) 0.05   (h) 0.5

(i) 0.253   (j) 0.331   (k) 0.881   (l) 0.457

3. Convert these percentages into fractions:

(a) 20%   (b) 65%   (c) 32%   (d) 75%

(e) 45%   (f) 80%   (g) 57%   (h) 37.5%

(i) 33%   (j) 95%   (k) 42%   (l) 87.5%

4. Convert these fractions into percentages:

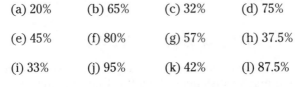

(a) $^{63}\!/_{100}$ (b) $^{43}\!/_{50}$ (c) $^{1}\!/_{25}$ (d) $^{3}\!/_{10}$

(e) ¼   (f) ⅗   (g) ½   (h) ⅝

(i) $^{12}\!/_{5}$ (j) $^{9}\!/_{20}$ (k) $^{1}\!/_{25}$ (l) ¾

5. What is 50% as (a) a fraction and (b) a decimal?

6. What is 0.05 as (a) a percentage and (b) a fraction?

7. What is ⅗ as (a) a percentage and (b) a decimal?

8. Alex ate 20% of a cake. Blake ate a quarter of an identical cake. Which of the two ate more cake?

9. Put the following into order, smallest first: ½; 2%; 0.2.

10. One shop offers 15% off all prices; another offers reductions of one-fifth of the full price. Which sale is better?

Splitting Things Up

Converting between percentages and fractions and decimals is all well and good, but it's not what you normally want to do with percentages. The thing you're most likely to need to do in real life is to work out a certain percentage of a number or, possibly, find out what percentage one number is of another (for example, answering 'what is 35 as a percentage of 70?').

In this section, I show you two ways to do these questions: one using the Table of Joy, which I explain in more detail in Chapter 7, and another, more traditional method using fractions.

Picking a percentage

Suppose you want to find what percentage one number is of another; for instance, it might be that 12 of the 30 people in your keep-fit class are male, and somebody asks you what percentage of the total that is. Here's how you work out that sum using the Table of Joy:

1. **Draw out a noughts and crosses grid.** Look at the one shown in Figure 8-1(a).

2. **Label the two right columns as 'People' and 'Percentage', and the two lower rows as 'Males' and 'Total'.**

3. **Fill in the numbers you know.** The total percentage is always 100, so write 100 in the bottom-right cell (percentage/total). There are 30 people in total, so put 30 in the bottom-middle square (people/total). And

there are 12 males in the class, so put 12 in the middle-middle square (people/males); see Figure 8-1(a). The middle-right square (people/percentage) is what you're trying to work out.

4. **Shade the grid like a chess board and write down the Table of Joy sum.** Times together the two numbers on the same colour, and divide by the other number. Here, it's $12 \times 100 \div 30$; see Figure 8-1(b).

5. **Work out the sum to get your answer.** It's $1200 \div 30$, which is 40%, as shown in Figure 8-1(c).

	People	Percentage
Males		
Total		

a)

	People	Percentage
Males	12	
Total	30	100

b)

$$\frac{12 \times 100}{30} = 40$$

c)

Figure 8-1: Finding a percentage with the Table of Joy: (a) Creating the grid; (b) Shading the grid and entering the numbers; (c) Working out the sum.

The traditional way of working this kind of sum out can be quite easy, but needs some thought as well. Follow these steps:

1. **Work out what fraction the problem represents by writing the 'part' number over the 'whole' number.** In this case, you get $\frac{12}{30}$.

2. **Cancel your fraction down.** For this example, you get $\frac{2}{5}$.

3. **Convert this number into a percentage using one of the recipes from the previous section.** So, $\frac{2}{5}$ is the same as 40%.

Picking a number

Working out what number is a certain percentage of another number is really straightforward with the Table of Joy. Here's the recipe to work out 35% of 20:

1. **Draw out a noughts and crosses grid.**

2. **Label the two right-hand columns as 'Number' and 'Percentage', and the two lower rows as 'Part' and 'Total'.** See Figure 8-2(a).

3. **Fill in the numbers you know.** The total percentage is always 100, which goes in the bottom-right square here. The part-percentage you're trying to find is 35, which goes in the middle-right square, and the total number is 20, which goes in the bottom-middle.

4. **Shade in the grid like a chessboard and write down the Table of Joy sum.** Times together the two numbers on the same-coloured squares and divide by the other number. For this one, the sum is $35 \times 20 \div 100$. See Figure 8-2(b).

5. **Work out the sum to get your answer.** As shown in Figure 8-2(c), it's $700 \div 100 = 7$.

	Number	Percentage
Part		
Total		

a)

	Number	Percentage
Part		35
Total	20	100

b)

$$\frac{35 \times 20}{100} = 7$$

c)

Figure 8-2: Finding one number as a percentage of another: (a) Drawing and labelling the grid; (b) Shading the grid and entering the numbers; (c) Working out the sum.

Of course, you can work out numbers as a percentage of other numbers in the old-fashioned way as well, if you prefer. Here's the trick:

1. **Convert the percentage you want into a fraction.** For this sum, it's 35/100, or 7/20.

2. **Work out this fraction of the number, using the recipe in 'Finding a Fraction of a Number' in Chapter 5.** So, ⅟₂₀ of 20 is 7.

Attempting some splitting things up questions

1. Work out:

(a) 10% of 70   (b) 20% of 85   (c) 50% of 84   (d) 25% of 24
(e) 90% of 250   (f) 75% of 64   (g) 70% of 30   (h) 22% of 150
(i) 80% of 120   (j) 40% of 45   (k) 20% of 15   (l) 13% of 300

2. Work out as a decimal:

(a) 30% of 35   (b) 10% of £15.60   (c) 63% of 20
(d) 16% of £60   (e) 24% of 10   (f) 90% of 12

3. Work out as a fraction:

(a) 50% of 1   (b) 20% of 4   (c) 25% of 3
(d) 90% of 7   (e) 60% of 12   (f) 75% of 3

4. What percentage is:

(a) 35 out of 50?   (b) 20 out of 25?   (c) 19 out of 20?

(d) 300 out of 400?   (e) 60 out of 75?   (f) 134 out of 200?

(g) 70 out of 80?   (h) 37 out of 74?   (i) 39 out of 52?

(j) £1.60 of £8?   (k) £1.50 of £15?   (l) 20p out of 50p?

5. Chris got 35 points out of 40 in his exam. Daljit got 44 points out of 50 in his. Which of them had the better percentage mark?

6. Mr Pink wants to add a 12% tip to his restaurant bill of £15. How much tip does he leave?

7. Fifteen per cent of the students in a school are left-handed. If the school has 500 people, how many are *not* left-handed?

8. Work out:

(a) 40% of 50   (b) 20% of 10   (c) 90% of 40
(d) 10 out of 25 as a percentage
(e) 9 out of 20 as a percentage
(f) 96 out of 200 as a percentage
(g) 35% of 30   (h) 45% of 12   (i) 17% of 10
(j) 8.5 out of 10 as a percentage
(k) 6.4 out of 8 as a percentage
(l) 12 out of 16 as a percentage

Calculating Change

When you're comfortable with working out the answer to a percentage question (and if you're not, I suggest you go back and attempt or redo the problems from earlier in the chapter), doing real-life percentage questions is only a small step further on. Percentages come up in many real-life situations, including tax, tips, sales, price increases, deposit schemes and interest at the bank. In this section, I take you through all of these uses of percentages so that you can race through the questions at the end!

Adding on tax and tips

One of the commonest places you see percentages (other than in maths questions, of course) is associated with prices. Often a sign in a shop says something like '20% off', but you also see things like a computer costing £350 plus 20% VAT.

Another situation in which you might see a percentage is when giving a tip. In many countries leaving a small amount of extra money for the waiter or waitress at the end of a meal is polite, especially if they've provided good service. How much you leave is up to you, of course, but in some countries it's almost mandatory to leave a certain percentage of the price of the meal (usually 10 to 20 per cent) as a tip. Being able to work out this kind of sum on the fly is very useful – it can save you from tipping too much, or from upsetting the people preparing your food.

If you want to work out the total cost of something including tax or a tip – let's say you want to add a 15% tip to a £30 meal – here's what you do:

1. **Work out what percentage you want to add.** In this case, it's 15%.

2. **Work out that percentage of the total.** So, 15% of 30 is £4.50.

3. **Add this amount on to the original number.** You pay a total of £34.50.

You can easily figure out how big a percentage increase or decrease is – such as 'my season ticket used to cost £180, but now it costs £189; what is the percentage increase?' – using the following recipe:

1. **Work out which of the numbers is the original or full amount.** Here, the original amount is £180.

2. **Find the difference between the two numbers.** This is the increase or decrease. Here, it's a £9 increase.

3. **Work out what percentage of the original number the increase or decrease is.** So, £9 out of £180 is 5%, which is the answer.

Taking away tax

Percentages – as far as basic maths is concerned – are pretty simple. You don't have to deal with them misbehaving, but I do want you to know that they can misbehave. The classic example is taking off tax after you've added it.

For example, if you buy a £100 computer and pay an extra 20% VAT, your total bill comes to £120. However, if you buy a computer for £120 including VAT and take 20% off, you end up with the pre-tax price being £96, not £100! Something is wrong here.

Well, it hopefully isn't a huge surprise that 20% of 100 and 20% of 120 are different. The problem is, the tax is always calculated on the *original* price (with this computer, £100), not on the after-tax price. The moral of the story: you can't get back to the pre-tax price by taking away the tax percentage. Luckily, you don't have to worry about this unless you're doing invoices or studying further than this book will take you.

Buying in a sale

I can't remember the last time I went into a shop and didn't see a sign or a sticker offering a percentage off some prices. Most of the time, the shops are counting on you not to do the sum, but to say 'ooh, that's cheap!' and buy more stuff than you otherwise would. But not any more! From now on, you'll get your pen and paper out and follow the recipe below for working out sales prices so you can see what you're spending:

1. **Work out what percentage you want to take off.**

2. **Find that percentage of the original price.**

3. **Take your answer from Step 2 away from the original price.**

For instance, if your £40 shirt was in a 20% sale, you'd work out 20% of £40 (which is £8) and take the answer away from £40. Your shirt would cost £40 – £8 = £32.

Attempting some taxes, tips and sales questions

1. If you have to pay VAT of 20% on the following items, what total price do you pay for each (including VAT):

(a) A £60 pair of shoes?   (b) A £450 computer?
(c) A £130 mobile phone?

2. What would be the cost of the following meals:

(a) A £14 meal with a 10% tip?   (b) A £56 meal with a 15% tip?
(c) A £30 meal with a 12% tip?

3. What would the following items cost:

(a) A £17 DVD in a 10% off sale?
(b) A £45 train ticket with a 30% off travel card?
(c) A £220 robotics kit in a 25% off sale?

4. Find the percentage increase or decrease if:

(a) A rucksack's price changes from £40 to £34.
(b) The number of pupils at a school increases from 700 students to 742.
(c) The price of a map changes from £8 to £10.

Working out interest

If you invest money in a bank account or something similar, you normally receive a certain amount of *interest* every day, month or year. This is almost always a percentage of the amount you invested, and depends on your bank, the type of account and exactly how disastrously the economy is performing at any given time.

As I write this book, my bank is offering around 4 per cent interest every year on ISAs, and a feeble 0.05 per cent on current accounts. What this means is, for every £100 I invest in the ISA for a year, I'll receive £4; if I put the money in my current account instead, they'll pay me . . . 5p. I'll try not to spend it all at once.

Borrowing money works just the same way, although it's you who has to pay the interest! If you borrow £1,000 at 9 per cent interest, you have to pay an extra £90 after the first year. (See the nearby 'Compounding interest' sidebar if you're interested in what happens after that!)

Here's how you work out the interest on a sum of money (you'll probably recognise the recipe):

1. **Work out what percentage you're working with.**

2. **Find that percentage of the original amount of money.**

3. **Decide whether the question is asking for just the interest, or the *total balance*, that is, how much money there is altogether.**

4. **To answer a question asking for the *interest*, you give the amount from Step 2. If it asks for the *total balance*, you need to add the original amount to the interest.**

Compounding interest

You may be wondering, what happens after the first year? Unfortunately, that's where things get complicated. Fortunately, these are not sums you have to worry about in basic maths, but it's worth taking a quick look so that you can see what lies ahead.

Here's the problem: if you invest £1,000 at 10 per cent interest, say, after one year you have £1,100. If you don't think very hard about it, you might expect to have £1,200 after 2 years, £1,300 after 3 years and so on. That system is called *simple interest* and is a poor shadow of what you can expect in real life – no matter how easy the sums are.

Instead, in the second year, you also get interest on the interest you've already earned! This system is called *compound interest*, and is the one used by all banks. You earn 10 per cent of £1,100, which is £110, so you end up with £1,210 (instead of £1,200). The following year, you earn 10 per cent of the £1,210, which is £121, so after three years you have £1,331. That difference might not seem like much, but it really adds up over time; in 20 years, your investment would be worth nearly £7,000 – under simple interest, it would only be worth £3,000.

Dealing with deposits and payment plans

The last percentages topic you need to know about is dealing with deposits. This type of question is a bit more complicated than the other things in this chapter, but you can break the problem down into smaller, easier chunks.

When you're dealing with complicated questions, keeping your work neat and organised is really important. You'd be surprised by how easy it is to make extra mistakes when you try to cram things into too small a space, or write things at random all over the page!

A typical payment plan question looks something like this: 'Joe agrees to buy a $6,000 car on a payment plan. The dealership asks for a 20% deposit, followed by 12 monthly payments. How much is each payment?' You tackle this problem by following these steps:

1. **Work out how much the deposit is.** In this case, it would be 20% of £6,000, which is £1,200.

2. **Take this away from the full cost of the car.** Doing so gives you a *balance* of £4,800.

3. **Divide the balance by the number of payments.** In this case, you work out £4,800 ÷ 12 = £400, which is the amount of each payment.

You can always check your work by working backwards to make sure you get to the original price!

Just for reference, here's how to find the percentage deposit if you know everything else:

1. **Multiply the size of each payment by the number of payments.**

2. **Take your answer from Step 1 away from the total cost.** This is the deposit.

3. **Work out what percentage the deposit is of the total cost.**

If you want to know how many payments you need to make, follow these steps:

1. **Work out the deposit by finding the right percentage of the total cost.**

2. **Find the balance by taking the deposit away from the total cost.**

3. **Divide the balance by the size of each payment.** This is the number of payments you need.

Attempting some interest and deposit questions

1. How much interest would you receive if you invested (for one year):

(a) £1,500 at 3% interest?   (b) £10,000 at 1.5% interest?

(c) £5,000 at 0.5% interest? (d) £120 at 2% interest?

2. How much interest would you pay on a one-year loan of:

(a) £5,000 at 15% interest?   (b) £500 at 12% interest?

(c) £15,000 at 8% interest?   (d) £900 at 9% interest?

3. How much money would you have in your account after one year if you invested:

(a) £1,000 at 2% interest?   (b) £1,200 at 1% interest?

(c) £3,000 at 2.5% interest?   (d) £1,500 at 5% interest?

4. What would your payments be if you bought:

(a) a £4,000 car, paid a 10% deposit and made 12 payments?

(b) a £650 sofa, paid a 20% deposit and made 8 payments?

(c) a £900 computer, paid a 15% deposit and made 15 payments?

5. What percentage deposit would you have to pay if you bought:

(a) A £1,200 painting, and made 12 payments of £90?

(b) £540 of furniture, and made 12 payments of £39.60?

(c) A £2,700 holiday, and made 15 payments of £153?

6. How many payments would you have to make if you bought:

(a) A caravan for £5,000, paid a 10% deposit and then made payments of £90?

(b) A fridge for £300, paid a 50% deposit and then made payments of £50?

(c) A magnificent coffee machine for £450, paid a 20% deposit and then made payments of £40?

Finding the right equation

In a multiple choice test, you may also be asked to suggest an equation to use to find out the answer to a percentage problem. This kind of sum can be tricky, because there's usually more than one way to work out a sum and the way you or I would do it isn't necessarily the way the examiner would do it. Because so many variations exist, all I can do is offer you a few tips to help you decide which one might be right:

- ✔ Write down the sum you would do and see which of the answers given is closest.

- ✔ Pay special attention to the numbers common to your way and the ways the question offers. Which ones are multiplied together? Which ones come after a divide sign?

- ✔ Think about which options are obviously wrong.

- ✔ If you can't see which one is right, try working out the answer your way, and then working out the sums. (This process is time-consuming . . . but it works!)

Working through Review Questions

1. A man is on a 2,500 calories per day diet. His breakfast contains 300 calories. What percentage of his daily calorie allowance comes from his breakfast?

2. The owner of a bike shop sells 24 women's bikes and 36 men's bikes in a week. What percentage of the bikes he sold were men's bikes?

3. On 1 January, a hill-walker walks a total of 12 miles. On 1 August, she walks a total of 19.5 miles. By what percentage has her mileage increased?

4. A salesperson's basic rate of pay of £15 per hour is increased by 4%. Which of these calculations gives his new pay rate?

A. $15 \times 4 \div 100$

B. $15 + 4 \times 100$

C. $15 \times 100 \div 4$

D. $15 \times 4 \div 100 + 15$

5. Quick-fire round! Find:

(a) 40% of 15   (b) 45 out of 75 as a percentage
(c) 95% of 80   (d) 40 increased by 20%
(e) 15% off of £85   (f) the percentage increase from 36 to 45

Checking Your Answers

For more practice with percentages, you can check out some of the questions in Chapter 15, which deals with graphs and tables. Many sums involving graphs also ask about percentages.

Converting percentages questions

1. (a) 0.03   (b) 0.7   (c) 0.47   (d) 0.98
   (e) 0.53   (f) 0.09   (g) 0.1   (h) 0.15
   (i) 0.4   (j) 0.33   (k) 0.42   (l) 0.95

2. (a) 32%   (b) 88%   (c) 7%   (d) 62%
   (e) 60%   (f) 38%   (g) 5%   (h) 50%
   (i) 25.3%   (j) 33.1%   (k) 88.1%   (l) 45.7%

3. (a) $\frac{1}{5}$ (b) $\frac{13}{20}$ (c) $\frac{9}{25}$ (d) $\frac{3}{4}$
   (e) $\frac{9}{20}$ (f) $\frac{4}{5}$ (g) $\frac{57}{100}$ (h) $\frac{3}{8}$
   (i) $\frac{33}{100}$ (j) $\frac{19}{20}$ (k) $\frac{21}{50}$ (l) $\frac{7}{8}$

4. (a) 63%   (b) 86%   (c) 28%   (d) 30%
   (e) 25%   (f) 60%   (g) 150%   (h) 62.5%
   (i) 240%   (j) 45%   (k) 16%   (l) 75%

5. (a) 1/2 (b) 0.5

6. (a) 5% (b) 1/20

7. (a) 40% (b) 0.4

8. Blake. Alex ate 20% of the cake, and Blake 25%.

9. 2%, 0.2, ½. 0.2 is 20% and ½ is 50%.

10. The one-fifth off sale. ⅕ is the same as 20%, which is a better sale than 15%.

Splitting things up questions

1. (a) 7	(b) 17	(c) 42	(d) 6
(e) 225	(f) 48	(g) 21	(h) 33
(i) 96	(j) 18	(k) 3	(l) 39

2. (a) 10.5	(b) £1.56	(c) 12.6	(d) 9.6
(e) 2.4	(f) 10.8		

3. (a) ½	(b) ⅘	(c) ¾	(d) 6 3⁄10
(e) 7 ⅕	(f) 2 ¼		

4. (a) 70%	(b) 80%	(c) 95%	(d) 75%
(e) 80%	(f) 67%	(g) 87.5%	(h) 50%
(i) 75%	(j) 20%	(k) 10%	(l) 40%

5. Chris got 35 × 100 ÷ 40 = 87.5%. Daljit got 44 × 100 / 50 = 88%. Daljit did better – but not by much!

6. £15 × 12 ÷ 100 = £1.80

7. 85 × 500 ÷ 100 = 425. You could also work out the number of left-handed students (15 × 500 ÷ 100 = 75) and take your answer away from 500 to get 425.

8. (a) 20	(b) 2	(c) 36	(d) 40%
(e) 45%	(f) 48%	(g) 10.5	(h) 5.4
(i) 1.7	(j) 85%	(k) 80%	(l) 75%

Taxes, tipping and sales questions

1. (a) £72   (b) £540   (c) £156

2. (a) £15.40   (b) £64.40   (c) £33.60

3. (a) £15.30   (b) £31.50   (c) £165

4. (a) 15% decrease   (b) 6% increase   (c) 25% increase

Interest and deposit questions

1. (a) £45   (b) £150   (c) £25   (d) £2.40

2. (a) £750   (b) £60   (c) £1,200   (d) £81

3. (a) £1,020   (b) £1,212   (c) £3,075   (d) £1,575

4. (a) £300   (b) £65   (c) £51

5. (a) 10%   (b) 12%   (c) 15%

6. (a) 50   (b) 3   (c) 9.

Review questions

1. 12%   2. 60%   3. 62.5%   4. (d)

5. (a) 6   (b) 60%   (c) 76   (d) 48
    (e) 68   (f) 25%

Part III
Real-life Maths

"All right – I read the weather forecast incorrectly and 30° Centigrade is certainly a <u>bit</u> warmer than 30° Fahrenheit."

In this part . . .

Finally! Something you can apply directly to your day-to-day life!

This part is all about things you encounter in reality – time and money, temperature, weights and measures – and shows you what you can work out quickly and easily.

Chapter 9

Clocking Time

. . . .

In This Chapter

▶ Telling the time

▶ Keeping to schedule

▶ Getting to grips with timetables

. . . .

I could rant for days on end about how stupid our system of time is. There's really no sensible reason for it. We have, for instance:

- ✔ 12 months in a year
- ✔ A variable number of days in each month
- ✔ Seven days in a week (so there are four and a bit weeks in most months, and 52 and a bit weeks in a year)
- ✔ 24 hours in a day
- ✔ 60 minutes in an hour
- ✔ 60 seconds in a minute

None of those numbers are particularly easy to work with, and for no good reason other than 'that's the way it's always been.' It'd be much nicer if everything was in powers of ten – it would make all of the sums in this chapter much easier! (The only time number that makes any sense at all is having 365 or 366 days in the year, because that's just about exactly how long it takes to travel around the sun.)

Phew! Rant over. Unfortunately, it seems unlikely that time will be decimalised in the near future, so you need to know the numbers from that list.

In this chapter, I show you how to adapt the sums you know and love to work with our silly system for measuring time, so that you can catch the right train and don't burn the lasagne.

Telling the Time

Figure 9-1 shows two kinds of clock – on the left, an *analogue* clock with hands and, on the right, a *digital* clock with numbers.

Figure 9-1: An analogue clock (left) and a digital clock (right).

In this section, I show you how to read both kinds of clocks in case you don't already know. If you do already know, you can either skim through this section feeling really smart, or you can jump straight to the 'Starting and Ending on Time' section to get down to the nitty-gritty.

Ticking along like clockwork

A *time* is normally made up of two parts: the *hour*, which is a number between one and 12, the *minute*, which is a number between 0 and 59, and two letters – either 'a.m.', which means morning, or 'p.m.', which means afternoon. (You sometimes see it without the a.m. or p.m. – that means the 24-hour clock or 'military time' is being used; have a look at the section on 'Marching to military time' later in this chapter.)

By convention, midday (12 o'clock lunchtime) is defined as being in the afternoon, so it's written as 12:00 p.m. Midnight is written as 12:00 a.m. Like with everything time-related, no particular reason for this exists, it just is.

Telling the time on a digital clock is pretty easy: you see a dot (.) or a colon (:) splitting a number into two parts. I prefer using colons - The first part is the hours and the second part is the minutes. Thus, the clock on the right of Figure 9-1 shows 50 minutes past one o'clock. (You may also see a third part, the seconds, but you're unlikely to have to worry about those.)

Telling the time on an analogue clock with hands is a bit more difficult. Here's a brief recipe:

1. **Look for the bigger hand and find the big hand going the same way in Figure 9-2.** The number underneath the picture is the number of minutes past the hour. Write this down after a colon (:). If your minute hand isn't pointing in a nice direction, work out which minutes it's between and make a sensible guess as to where it is. (For instance, if it's between 15 and 30, the time is likely to be either 20 or 25 – which does it look closer to?)

2. **Look at where the smaller hand is, and work out which of the directions in Figure 9-3 is the closest to it. If it's in-between, again make a sensible guess.**

3. **If your answer in Step 1 was less than or equal to 30, write your answer from Step 2 before your colon and you're finished.**

4. **If your answer in Step 2 was more than 30, take one away from your answer in Step 2 and write your answer in front of the colon.**

5. **If you end up with zero in front of the colon, you need to tut, cross it out and replace it with 12.**

There! Simple.

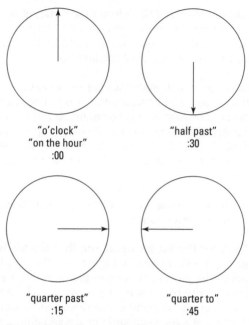

Figure 9-2: Working out the minute hands.

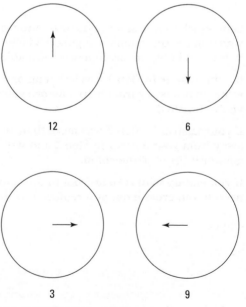

Figure 9-3: Working out the hour hands.

Marching to military time

A problem is evident with normal times, which is that if I say 'I'll see you at nine o'clock', you don't know whether I mean the morning or evening – there are two nine o'clocks in the day.

One solution to this problem is to make sure you always say 'a.m.' or 'p.m.' with the time. However, another way to differentiate is to use the *24 hour clock* or, because it was originally used mainly by the army, *military time*.

Here's how you convert an ordinary time to military time:

1. **If your time is in the morning (and the hour is between 1 and 9), write zero, then the hour, then the minutes with no colon.** So, 9:45a.m. becomes 0945.

2. **If your time is around lunchtime and the hour is between 10 and 12, just write the hours followed by the minutes with no colon.** 11:30 a.m. becomes 1130.

3. **If your time is in the afternoon or evening and the hour is between 1 and 11, add 12 to the number of hours, then write this number followed by the minutes, with no colon.** So, 2:15pm becomes 1415.

4. **If your time is in the middle of the night and starts with 12, write 00 followed by the number of minutes.** So, 12:20 a.m. becomes 0020.

Translating time from military time to normal time is easy as well:

1. **Split the number into two pairs.** The first two digits are the hours number and the second pair are the minutes number.

2. **Write down a colon followed by the minutes number.** Make sure you leave enough space before it for the hours.

3. **If the hours number is between 00 and 11, it's in the morning, so write 'a.m.' after the minutes number.** Otherwise, write 'p.m.'

4. **If the hours number is 00, write down '12' before the colon in Step 2.**

5. If the hours number is between 01 and 12, write it down before the colon, but if there's a zero at the start of the number you don't have to write it.

6. If the hours number is between 13 and 23, take away 12 from it and write it down before the colon.

Attempting some clock questions

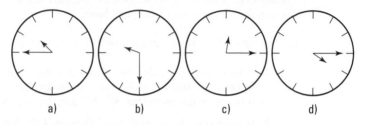

a) b) c) d)

1. What time is it on clocks a), b), c) and d) in the above figure, in normal time? (You can't tell if they mean morning or afternoon, so don't worry about writing a.m. or p.m. for this question.)

2. Using military time, what time is it on clocks a), b), c) and d) in the above figure if you're reading them in the morning?

3. Using military time, what time is it on clocks a), b), c) and d) in the above figure if you're reading them in the afternoon or evening?

4. Draw clock faces showing the following times:

(a) 9:00 a.m. (b) 3:15 p.m. (c) 7:30 p.m. (d) 4:20 a.m.
(e) 1300   (f) 1545   (g) 0330   (h) 0030

5. What are the following military times in normal time?

(a) 1045   (b) 1440   (c) 2359   (d) 0410

6. What are the following normal times in military time?

(a) 3:45 a.m. (b) 6:50 p.m. (c) 9:30 a.m. (d) 12:00 a.m.

Starting and Ending on Time

Probably the commonest time question you see in exams –
and the most likely thing you have to think about in relation to
time in real life – involves start times, end times and durations.

This kind of question is generally posed in one of three ways:

> ✔ You know when something starts and how long it takes,
> but not when it finishes.
>
> ✔ You know when something starts and when it ends, but
> not how long it takes.
>
> ✔ You know how long something takes and when it fin-
> ishes, but not when it starts.

In each case, you need to figure out the thing you don't know.
In this section, I show you how to work out the answer to
each type of question.

Doing sums with time has one really tricky element: hours
comprise only 60 minutes rather than 100. That means you
can't just do adding and taking away as normal with times –
you have to be very careful about how you break down such
questions to work them out. I show you how in the following
sections.

Working out starting times

To work out when an event started, if you know how long it
took and when it ended, follow these steps:

1. **Split both times into hours and minutes.**

2. **If the number of end time minutes is *smaller* than the
   length minutes, add 60 to the end time minutes and
   take away one from the number of end time hours.**

3. **Take away the length minutes from the end time minutes and write down the number in front of a colon.**

4. **If the number of end time hours is *smaller* than the
   length minutes, add 12 to the end time hours, and
   switch from a.m. to p.m.**

> **5. Take away the length hours from the end time hours. Write the resulting number in front of the minutes in Step 3 and you have your answer.**

So, for example, if your football match ends at 8:30 p.m. and lasts 1 hour and 45 minutes (including half time, of course!), when is kick-off? The end time is 8 hours and 30 minutes; the duration is 1 hour and 45 minutes, but 45 is too big to take away from 30. So I add 60 to 30 (to get 90) and take one from the hours – saying that the match ends at 90 minutes past 7. Now I can take away the 45 minutes (to get 45) and take away 1 hour of duration from the 7 in the end time to get 6. The match started at 6:45pm.

Lasting the distance

Finding how long an event lasted is almost exactly the same process as for finding out when it started! For the sake of completeness, to work out when an event started, if you know how long it took and when it ended, follow these steps:

> **1. Split both times into hours and minutes.**

> **2. If the number of end time minutes is *smaller* than the start time minutes, add 60 to the end time minutes and take away one from the number of end time hours.**

> **3. Take away the start time minutes from the end time minutes and write down the number in front of a colon.**

> **4. If the number of end time hours is *smaller* than the start time minutes, add 12 to the end time hours.**

> **5. Take away the start time hours from the end time hours. Write the resulting number in front of the minutes in Step 3 and you have your answer.**

So, if my train leaves at 6:45 a.m. and arrives at 10:20 a.m., how long does my journey take? I can't do 20 – 45, so I need to turn 10:20 into 80 minutes past 9. Then I do 80 – 45 = 35 to get the minutes, and 9 – 6 = 3 to get the hours. My journey takes 3 hours and 35 minutes.

Calculating ending times

Finding out when something finishes, if you know the start time and how long it takes, is slightly the odd one out in this section. Instead of taking the times away, you have to add them on – which does actually make sense if you stop to think about it.

Here's the recipe:

1. **Split both times into hours and minutes.**

2. **Add the start time minutes to the duration minutes.**

3. **If the answer is more than 60, take away 60 and write down the answer after a colon; also add one to the start time hours.**

4. **Add the start time hours and duration hours together. If the answer is more than 12, take away 12 and switch from a.m. to p.m. or vice versa.**

5. **Write the resulting number in front of the minutes in Step 3 and you have your answer.**

Attempting some scheduling questions

1. At what time did the following events start?

(a) A flight that lasted two hours and 15 minutes, and landed at 6:00 p.m.

(b) A cake that needed to be baked for 45 minutes, and to be ready at 8:30 p.m.

(c) A film that ended at 10:00 p.m. and lasted for one hour and 50 minutes.

(d) A series of 10 meetings, each lasting 20 minutes, which need to be finished by 5:00 p.m. (Hint: work out how long the meetings take altogether and turn that number of minutes into hours.)

2. How long did the following events take?

(a) A charity race I started at 9:30 a.m. and finished at 10:45 a.m. (You try running 100 metres dressed as a hare!)

(b) A train journey that left at 5:30 p.m. and arrived at 6:05 p.m.

(c) A concert that started at 7:30 p.m. and ended at 10:15 p.m.

(d) A film that started at 3:20 p.m. and finished at 5:05 p.m.

3. At what time would the following events finish?

(a) A meeting that starts at 10:30 a.m. and goes on for 45 minutes.

(b) A class that lasts one hour and 30 minutes, beginning at 6:15 pm.

(c) A car journey that lasts two and a half hours, leaving at 11:00 a.m.

(d) A training session that lasts one hour and 15 minutes, beginning at 6:45 p.m.

Following Timetables

Here's one of the commonest real-life situations in which you may need to work out time sums in your head: you need to catch a bus or a train or a plane somewhere, you need to be there by a particular time, and you need to figure out when you need to be at the bus stop (or train station or airport).

Personally, I find timetables quite intimidating, and I'm not usually easily frightened by huge sheets of numbers. In fairness, many travel companies have worked hard in the last few years to make timetables easier to read and understand (especially online) but you still need to be able to read the big ones once in a while – and in the exam, if nothing else.

If you find yourself becoming overwhelmed by the sheer size of a timetable, take a deep breath, stand up or sit up straight and remind yourself that you can do this!

Taking your time

Cardiff	0800	0900	1000	1100
Hereford	0900	1000	1100	1200
Crewe	1030	1130	1230	1330
Manchester	1115	1215	1315	1415

Figure 9-4: Looking at a timetable.

I'm sure you've seen a timetable at a bus stop like the one shown in Figure 9-4. Down the side, you have all the stops the bus makes and, each column shows a different bus that runs the same route. In the grid, you've got the time each bus is supposed to leave each stop.

Reading the time the bus or train is supposed to be at each stop is easy: you just read across from the name of the stop until you hit the time in the same column as the bus (or train) you're interested in.

Timetable questions are very similar to all the other time questions: you just have the extra step at the beginning in which you find the times you're interested in and then you work out the sum as normal.

Attempting some timetable questions

Liverpool	1400	1600	1800	2000
Manchester	1500	1700	1900	2100
Stockport	1510	1710	1910	2110
Sheffield	1600	1800	2000	2200

1. The above figure shows a timetable for several trains between Liverpool and Sheffield.

(a) What time does the last train leave Manchester?

(b) How long does the 4:00 p.m. train take to get from Liverpool to Sheffield?

(c) If I want to arrive in Sheffield before 9:00 p.m., what's the latest train I can catch from Liverpool?

(d) If I want to arrive in Sheffield before 5:00 p.m., what's the latest train I can catch from Stockport?

Bournemouth	0600	0800	1000
Southampton	0700	0900	1100
Oxford	0910	1110	1310
Birmingham	1200	1400	1600
Carlisle	1610	1810	2010
Glasgow	1810	2010	2210

2. The above figure shows a bus timetable between Bournemouth and Glasgow.

(a) When does the first bus arrive in Birmingham?

(b) How long does it take the last bus of the day to go from Carlisle to Glasgow?

(c) If I want to be in Birmingham before 4:30 p.m., what time must I leave Oxford?

(d) How many buses arrive in Glasgow after 8:00 p.m.?

Speeding Along

You may need to know how to work out speed, distance and time sums. Only three types of sum are involved here:

- ✔ To work out a *speed*, you divide the distance by the time.

- ✔ To work out a *distance* (in kilometres or miles), you multiply the speed (in kilometres per hour or miles per hour) by the time (in hours).

- ✔ To work out a *time*, you divide the distance by the speed.

You can remember the order in which you need to work things out using a formula triangle like the one shown in Figure 9-5. I also use the mnemonic 'sun dried tomatoes' to remember which letters go where, starting from the bottom-left and going over the top.

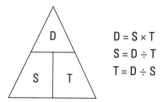

$$D = S \times T$$
$$S = D \div T$$
$$T = D \div S$$

Figure 9-5: Working out speed, distance and time using a formula triangle.

Ensuring your sum makes sense

Whenever you work out a speed–distance–time sum, make sure your answer makes sense. If you end up with a bike going at 900 kilometres per hour, you've probably made a mistake. If a plane journey works out to be half a mile, that doesn't seem very likely either.

The three commonest mistakes with this kind of sum are:

- ✔ Doing the wrong sum! Double check whether you're meant to times or divide.

- ✔ Using the wrong units. If your speed is in kilometres per hour, make sure your distance is in kilometres and your time in hours. Convert if you need to.

- ✔ Getting confused with the number of hours. It's very easy to think that 2 hours and 30 minutes is 2.3 hours, but it's not! It's two and a half hours, or 2.5 hours. Make sure that you write down the hours and minutes correctly

If your time is in minutes, thinking about how far you'd go in one minute and then multiplying that by 60 is a good idea. You're more likely to use this method in real life than in a test, though.

Attempting some speed questions

1. How long (in hours and/or minutes) does it take to travel:

(a) 80km at 40km/h?

(b) 100km at 60km/h?

(c) 400km at 80km/h?

(d) 20km at 60km/h?

(e) 90 miles at 60mph?

(f) 50 miles at 60mph?

(g) 100 miles at 30mph?

(h) 100 miles at 40mph?

2. How far would you travel if you travelled for:

(a) Two hours at 50km/h?

(b) Five hours at 70mph?

(c) Two and a half hours at 90km/h?

(d) Twenty minutes at 30km/h?

(e) Three hours at 60mph?

(f) Eight hours at 50mph?

(g) An hour and a quarter at 40mph?

(h) One hour and 40 minutes at 45mph?

3. What speed would you be travelling at if you covered:

(a) 80km in two hours?

(b) 90km in 3 hours?

(c) 45km in half an hour?

(d) 100km in 4 hours?

4. What speed would you be travelling at if you covered:

(a) One mile in one minute?

(b) One mile in two minutes?

(c) 50 miles in two hours?

(d) 30 miles in half an hour?

Working through Review Questions

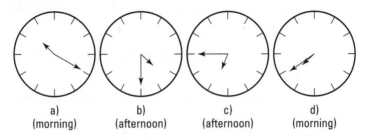

a)
(morning)

b)
(afternoon)

c)
(afternoon)

d)
(morning)

1. Look at clocks a), b), c) and d) in the above figure. What is the (normal) time on each of them?

2. Look at clocks a), b), c) and d) in the above figure. What is the military time shown on each of them?

3. What time is each of the following in normal time?

(a) 0125   (b) 1700   (c) 2310   (d) 2140

4. What time is each of the following in military time?

(a) 12:20 p.m.   (b) 9:40 p.m.   (c) 5:30 a.m.   (d) 11:59 p.m.

5. What time would the following events have started? (I know the times are funny; it's to do with time zones.)

(a) An American football match that ended at 2:00 a.m. and lasted for three and a quarter hours.

(b) A tennis match that ended at 6:00 a.m. and lasted for two and a half hours.

(c) A cricket match that ended at midday and lasted for five hours and 45 minutes.

(d) An athletics meeting that ended at 6:15 p.m. and lasted for eight and a half hours.

6. What time would my flight arrive if:

(a) It was supposed to land at 9:30 a.m. but was delayed by two and a half hours?

(b) It was meant to land at 6:00 a.m., but arrived 40 minutes early?

(c) It took off at 2:45 p.m. and the flight lasted an hour and a half?

(d) It took off at 7:45 p.m. and the flight lasted 3 hours and 50 minutes?

7. My ferry arrived at 5:45 p.m., an hour and a half late. The crossing was meant to take 2 and a quarter hours. When did the ferry start its journey?

8. When would the following events have started?

(a) An hour and a half-long meeting that ended at 1:00 p.m.

(b) A three-quarters of an hour interview that ended at 3:30 p.m.

(c) A four and a half hour journey that ended at 3:45 p.m.

(d) A series of 10 five-minute presentations that ended at 5:00 p.m.

9. How long (in hours and minutes) would each of these events take?

(a) A conference made up of six half-hour talks and two one-hour breaks.

(b) A concert made up of 25 three-minute-long songs.

(c) A training session consisting of 11 two-minute exercises and 10 breaks, each lasting half a minute.

(d) A museum tour that started at 10:30 a.m. and ended at 11:15 a.m.

10. When would each of these events begin?

(a) A train journey that lasted five hours and 55 minutes, arriving at 1:10 p.m.

(b) A work session that lasted 30 minutes and ended at 12:20 p.m.

(c) A parents' evening that consisted of 20 meetings, each lasting five minutes, and that ended at 8:00 p.m.

(d) A school day that ends at 4:00 p.m., consisting of five classes of one hour and ten minutes' duration, plus an hour for lunch and a 15-minute break in the morning.

11. A famous Tommy Cooper joke goes as follows: I backed a horse last week, at twenty to one. It came in at quarter past three. How long would this comical horse have taken to run the race?

Iverness	0800	1000	1200	1400	1600
Aviemore	0850	1050	1250	1450	1650
Pitlochry	1005	1105	1305	1505	1705
Perth	1050	1150	1350	1550	1750
Stirling	1150	1350	1550	1750	1950
Glasgow	1235	1435	1635	1835	2035

12. Looking at the timetable in the above figure:

(a) How long does it take to get from Inverness to Perth?

(b) If I want to be in Glasgow for 5:00 p.m., what's the latest bus I can catch from Pitlochry?

(c) If I arrive at Inverness at 12:15 p.m., how long must I wait for a train that will take me to Aviemore?

(d) Katherine lives two and a half hours from Inverness. She wants to catch the bus that gets her to Glasgow for 4:35 p.m. What's the latest she can leave home?

13. How far would I get if I travelled for:

(a) Two hours at 40km/h?   (b) Three hours at 60km/h?

(c) Eight hours at 30km/h?   (d) Six hours at 15km/h?

14. How long would it take me to get:

(a) 100 km at 25km/h?   (b) 90km at 30km/h?

(c) 135 km at 45km/h?   (d) 300km at 10km/h?

15. What speed would I need to travel to get:

(a) 50km in 2 hours?   (b) 400km in 8 hours?

(c) 90km in 3 hours?   (d) 350km in 5 hours?

16. How far would I get if I travelled:

(a) At 90km/h for an hour and a half?
(b) At 30km/h for three and a half hours?
(c) At 60km/h for one hour and 20 minutes?
(d) At 60km/h for one hour and 10 minutes?

17. How long (in hours and minutes) would it take me to get:

(a) 50km at 40 km/h?   (b) 90km at 60 km/h?
(c) 100km at 40 km/h?   (d) 495km at 45km/h?

Checking Your Answers

The commonest mistake in time questions is to work out sums as if there were 100 minutes in an hour! Make sure you consciously turn an hour into 60 minutes if you don't have enough minutes, or 60 minutes into an hour if you have too many!

Clock questions

1. (a) 10:45   (b) 9:30   (c) 12:15   (d) 4:15

2. (a) 1045   (b) 0930   (c) 0015   (d) 0415

Military times always have four digits, so you need to have a zero in front if your hour number is less than ten!

3. (a) 2245   (b) 2130   (c) 1215   (d) 1615

4.

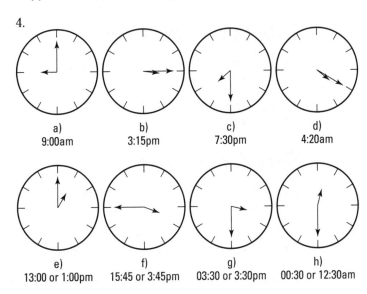

| a)
9:00am | b)
3:15pm | c)
7:30pm | d)
4:20am |

| e)
13:00 or 1:00pm | f)
15:45 or 3:45pm | g)
03:30 or 3:30pm | h)
00:30 or 12:30am |

5. (a) 10:45 a.m.   (b) 2:40 p.m.   (c) 11:59 p.m.   (d) 4:10 a.m.

6. (a) 0345   (b) 1850   (c) 0930   (d) 0000

Scheduling questions

1. (a) 6:00 p.m. – 2:15 = 3:45 p.m.
   (b) 8:30 p.m. – 0:45 = 7:45 p.m.
   (c) 10:00 p.m. – 1:50 = 8:10 p.m.
   (d) The meetings take a total of 200 minutes, or 3 hours 20.
   5:00 – 3:20 = 1:40 p.m.

2. (a) 10:45 a.m. – 9:30 a.m. = 1:15
   (b) 6:05 p.m. – 5:30 p.m. = 0:35
   (c) 10:15 – 7:30 = 2:45
   (d) 5:05 – 3:20 = 1:45

3. (a) 10:30 a.m. + 0:45 = 11:15 a.m.
   (b) 6:15 p.m. + 1:30 = 7:45 p.m.
   (c) 11:00 a.m. + 2:30 = 1:30 p.m.
   (d) 6:45 p.m. + 1:15 = 8:00 p.m.

Timetable questions

1. (a) 2100 or 9:00 p.m.   (b) 2 hours   (c) 1800 or 6:00 p.m.
   (d) 1510 or 3:10 p.m.

2. (a) 1200 or 12:00 p.m. (noon)   (b) 2 hours
   (c) 1310 or 1:10 p.m.   (d) two

Speed questions

1. (a) 2:00   (b) 1:40   (c) 5:00   (d) 0:20
   (e) 1:30   (f) 0:50   (g) 3:20   (h) 2:30

2. (a) 100km   (b) 350 miles (c) 225km   (d) 10km
   (e) 180 miles (f) 400 miles (g) 50 miles (h) 75 miles

3. (a) 40km/h   (b) 30km/h   (c) 90km/h   (d) 25km/h

4. (a) 60mph   (b) 30mph   (c) 25mph   (d) 60mph

Review questions

1. (a) 10:20 a.m. (b) 4:30 p.m. (c) 6:45 p.m. (d) 7:40 a.m.

2. (a) 1020   (b) 1630   (c) 1845   (d) 0740

3. (a) 1:25 a.m. (b) 5:00 p.m. (c) 11:10 p.m. (d) 9:40 p.m.

4. (a) 1220   (b) 2140   (c) 0530   (d) 2359

5. (a) 10:45 p.m. (b) 3:30 a.m. (c) 6:15 a.m. (d) 9:45 a.m.

6. (a) 12:00 p.m. (noon)   (b) 5:20 a.m.
   (c) 4:15 p.m.   (d) 11:35 p.m.

7. The ferry should have arrived at 4:15 p.m., so it would have left 2 hours 15 minutes before that, at 2:00 p.m..

8. (a) 11:30 a.m.   (b) 2:45 p.m.   (c) 11:15 a.m.
  (d) The presentations lasted a total of 50 minutes, so they started at 4:10 p.m.

9. (a) The talks lasted 3 hours and the breaks two hours, so five hours altogether
  (b) 25 × 3 = 75 minutes, making one hour and 15 minutes
  (c) The exercises last 22 minutes and the breaks 5 minutes, so 27 minutes altogether
  (d) 45 minutes

10. (a) 7:15 a.m.   (b) 11:50 a.m.
  (c) The meetings lasted 100 minutes, or 1:40; they started at 6:20 p.m.
  (d) The classes lasted 5:50, plus 1:15 of breaks, making 7:05 altogether; to end at 4:00 p.m., you'd have to start at 8:55 a.m.

11. From 12:40 to 3:15 is 2:35.

12. (a) 2 hours 50 minutes   (b) 1305 or 1:05 p.m.
  (c) 1 hour 45 minutes   (d) 9:30 a.m.

13. (a) 80km   (b) 180km   (c) 240km   (d) 90km

14. (a) 4 hours (b) 3 hours (c) 3 hours (d) 30 hours

15. (a) 25km/h (b) 50km/h (c) 30km/h (d) 70km/h

16. (a) 135km   (b) 105km   (c) 80km   (d) 70km

17. (a) 1:15   (b) 1:30   (c) 2:30   (d) 11 hours

Chapter 10

Counting the Cash: Dealing with Money

. . . .

In This Chapter

▶ Working with pounds and pence

▶ Figuring out different currency conversions

▶ Linking different methods together to solve more complicated sums

. . . .

*N*early all of the sums you consciously do in everyday life are to do with money. (Subconsciously, you do an awful lot of very complicated probability, calculus and game theory every time you so much as cross the road, but you already know how to do that). Instead, most of the sums an average person does on an average day include:

✔ Working out your budget

✔ Figuring out how much several items cost

✔ Dealing with complicated bargains

✔ Working out percentages for sales and interest

And so on. Even on a not-very-typical day, you may need to work out how much foreign currency you get for your British spending money or how much you have to pay each month for a complicated deposit scheme.

In this chapter, I show you how to do the calculations to deal with all of these. The good news is – you probably already know how to do most of them!

Calculating with Cash

Money sums work just like regular sums. If you're comfortable with decimals (check out Chapter 6 if you're not), much of this chapter will be a doddle for you. If you're not, then it could serve as a good introduction to decimals!

I'm sure you're familiar with money (you know, the paper and metal stuff in your purse or wallet), but I'm a mathematician and all mathematicians love stating the obvious in complicated ways. So, I'm thrilled to tell you that in the UK:

- ✔ Money is made up of *pounds* and *pence*.

- ✔ A *pound* is the same thing as 100 pence.

- ✔ You write the number of pounds after a £ symbol – ten pounds is £10.

- ✔ You write the number of pence before a p – so twenty pence is 20p.

- ✔ If you have a mixture of pounds and pence, you write the pounds, then a dot, then the pence, like this: £10.20 is ten pounds and twenty pence.

Bear in mind this irregular plural: one penny, two pence.

In this section, I talk you through the basics of doing money sums, how to set out your work out so the processes are easy and how to pick which sum you need to do.

Making sense of money sums

I can't stress this enough: *money sums work just like normal sums*. If you want to add two amounts of money together, you lay the numbers out just like a normal sum, add them up like normal and provide the result. No funny stuff. And it's the same with taking away.

If you want to multiply an amount of money by a number, guess what? You use exactly the same process as for multiplying normal numbers. And that stands for dividing, too.

The only mistake I see people make once in a while is mixing up pounds and pence. So, for instance, if they see the question:

A guitar costs £25 and a songbook costs £4.95. What does it cost to buy both together?

they may try to work out the sum using the process shown in Figure 10-1(a) – and get it wrong. What they need to do is set out the problem as shown in Figure 10-1(b), so they get the right answer. (The problem was simply writing the second number in the wrong place; it should have been two spaces to the left so that the 'pounds' column lines up.)

```
      2 5      ◄──── How not to do it!
   4. 9 5
               This would be £4.95 + 25p, not £25.
   5. 2 0
   1. 1


   2 5         ◄──── The right way!
   4. 9 5
               The 25 goes in front of the line.
   2 9. 9 5
```

Figure 10-1: Adding up money: (a) The wrong way; (b) The right way.

You can avoid making this kind of error by doing some estimation beforehand (estimating is covered in Chapter 4). So, the songbook is roughly £5, so the guitar and songbook together must be about £30. An answer of just over £5 is quite obviously wrong.

Lining up your dots

Lining up your dots when you're adding up or taking away money is *vital* so that you don't mix up pounds and pence (which was exactly the problem demonstrated in Figure 10-1).

Here's how you make sure that you set out an add or take away money sum properly, even if you don't have squared paper:

1. **Draw a vertical line that will be long enough to run through all of the numbers you'll be writing down.**

2. **If the number you're about to write down has a dot in it, write it so the dot lies on the line.**

3. **If the number you're about to write down is whole pounds, write it so that it ends just to the left of the line.**

4. **If the number you're about to write down is just pence, write it so that it starts just after the line.**

5. **Now do the sum as normal – and don't let the line distract you!**

Figure 10-2 demonstrates how to lay out the sum for adding up £24.50, 47p and £19.

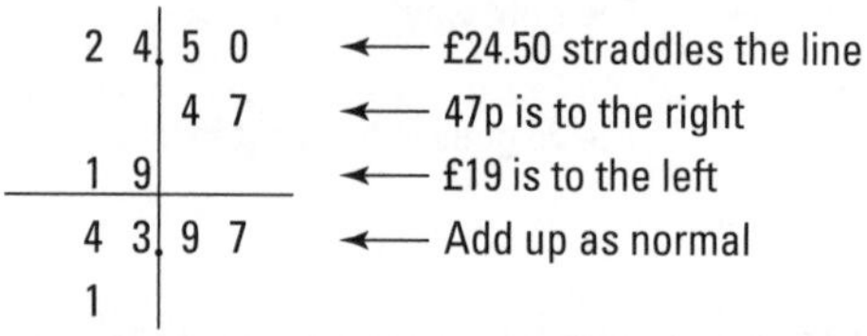

```
 2 4. 5 0    ◄──── £24.50 straddles the line
       4 7   ◄──── 47p is to the right
   1 9       ◄──── £19 is to the left
 4 3. 9 7    ◄──── Add up as normal
 1
```

Figure 10-2: Writing out a sum with a line through it.

Picking the right sum

Picking the right sum can be as much of an art as a science, but here are some tips to help you decide which sum is likely to be right:

✔ If you want to know the total cost of a number of differently-priced items, you need to add up the prices.

✔ If you want to know the total cost of several items that cost the same amount, you need to multiply the price by the number of items.

✔ If you want to know how much change you expect to get, or how much money you have left over, you need to take the price away from how much money you start with.

✔ If you want to split a sum of money evenly between people or items, you need to divide the money by the number of people or items.

✔ If you want to work out something to do with percentages, you need to follow the percentage recipes later in this chapter.

✔ If you want to do a calculation involving two different currencies, you need to follow the currency conversion methods described in the next section.

Attempting some cash calculation questions

1. Work out:

(a) £42.20 + £61.72 (b) £90.90 – £87.43 (c) £62.07 + 91p
(d) £57.20 – £34   (e) £75.36 + £85.87 (f) £91.80 – £40.56

2. Calculate:

(a) £2.20 × 6 (b) £5.80 × 3 (c) £35.50 × 4 (d) £68.30 ÷ 10
(e) £75 ÷ 5   (f) £12.60 ÷ 7

3. Tickets for an exhibition cost £7.50 for an adult and £3.90 for a child. How much would it cost to buy tickets for:

(a) Two adults?   (b) Two children?
(c) Two adults and two children?
(d) Three adults and a child?
(e) An adult and five children?

4. A lottery syndicate wins a prize of £18,000 to split between its members. How much does each winner receive if the syndicate has:

(a) Four members? (b) Twelve members? (c) Ten members?
(d) Five members? (e) Nine members?

5. I go into a shop with £25 in cash. How much change should I get if I buy:

(a) A bottle of wine for £8.99?
(b) A packet of chewing gum for 55p?
(c) Four big pots of yoghurt at £1.20 each?
(d) Everything in parts (a) to (c) at once?

Converting Currency

Different countries use different currencies. Many of the countries in Europe use the euro (€), which is made of 100 eurocents (c) America, Canada, Australia and New Zealand all use different kinds of dollar ($), each of which is made up of cents (c). Elsewhere in the world, you get all manner of currencies, from the dinar to the dong, and the rouble to the rupee, but they all work in the same way.

You can convert any currency to British money (or vice versa) using the Table of Joy (see Chapter 7 for a full description of how to use this marvellous grid), and if you know corresponding amounts in two different currencies, you can work out the exchange rate. In this section, I show you how to do both of these things!

Changing money

If you walk down any high street, you'll probably see a flashy display (like the one shown in Figure 10-3) in a travel agent's window covered in currencies and numbers, telling you how many euros, dollars, yen or roubles they'll give you in exchange for a pound.

£1 buys:

US Dollars:	$1.60
Euros:	€1.20
Australian Dollars:	$1.50
Canadian Dollars:	$1.60
Indian rupees:	Rps. 80
South African Rand:	ZAR12.20
New Zealand Dollars:	$1.90
Japanese Yen:	¥120

Figure 10-3: A travel agent's window display (simplified).

Places where you can change money – *bureaux de change* – usually list two different rates for each currency, one for buying and one for selling, so that they can make a tidy profit on any transaction. If you turned £100 into a different currency and then changed it straight back again, you'd end up with less money than you started out with! For simplicity's sake, I've only listed one rate for each currency in Figure 10-3.

This simplified figure is only helpful, however, if you're planning to take only one pound of spending money with you. However, even my super-stingy budget holidays tend to call for a little bit more spending power than that. So how do you change, say, £400 into US dollars? Well, one way is to use the Table of Joy, as I show in Figure 10-4 and describe in the following steps:

1. **Draw out a big noughts and crosses grid, leaving plenty of room for labels.**

2. **Label the columns.** In the top-middle and top-right squares list the currencies you want to exchange – here, '£' and '$' (or 'Pounds' and 'Dollars', if you're not into the whole brevity thing).

3. **Label the rows.** In the left-middle square, write 'Rate' and in the bottom left cell, write 'Money'.

4. **Fill in the numbers.** You know by looking at the board that £1 is the same as $1.60, so you put 1 in the £/rate square and 1.60 in the $/rate square. Also, you know that you have £400 in money, so you write 400 in the £/money square.

5. **Shade in the squares and write down the Table of Joy sum.** Here, the 1.60 and the 400 are both on unshaded squares, and the 1 on a shaded square, so the sum is $1.60 \times 400 \div 1$.

6. **Now work out the sum!** So, $1.6 \times 400 = \$640$.

	£	$
Rate	1	1.60
Money	400	

$$\frac{1.60 \times 400}{1} = 640$$

Figure 10-4: Converting currency using the Table of Joy.

This method also works the other way round: if you have $400 and you want to know how much that is in pounds, you can fill in the Table of Joy in the same way, except that you put the 400 in the bottom-right, $/money square. The sum works out to be 400 × 1 ÷ 1.6 = £250.

Finding the rate

Another possible question you may have to deal with in relation to currency is working out the exchange rate if you know the price of one thing in both currencies. For example, a DVD in an airport shop may be advertised as £15 or €18 and you – being a curious sort – would want to know the exchange rate.

You almost always give the exchange rate as 'one pound is worth . . .' a certain amount of foreign currency. Only if asked specifically would you say 'One euro is worth. . .' some number of pounds.

It's time to break out the Table of Joy again!

1. **Draw out a big noughts and crosses grid, leaving plenty of room for labels.**

2. **Label the columns.** In the top-middle and top-right squares list the currencies you want to exchange – here, pounds (£) and euros (€).

3. **Label the rows.** In the left-middle square, right 'Rate' and in the bottom-left cell, write 'Money'.

4. **Fill in the numbers.** You know that £15 is worth the same (in this shop) as €18, so write 15 in the £/money square and 18 in the €/money square. Also, you want to know how much one pound is worth, so write 1 in the £/rate cell.

5. **Shade in the squares and write down the Table of Joy sum.** Here, the 1 and the 18 are shaded, and the 15 unshaded. The sum is 1 × 18 ÷ 15.

6. **Now work out the sum!** So, 18 ÷ 15 = 1.2 – so £1 = €1.20.

If you're writing down an amount of money, you normally want to give it to two decimal places. If you only have one decimal place, as in this example, put a zero on the end – at the end of a number, after the decimal point, adding a zero doesn't change anything.

Attempting some currency questions

1. If one pound is worth €1.20, how many euros is each of the following worth?

(a) £10   (b) £30   (c) £5   (d) £35
(e) £120   (f) £55   (g) £64   (h) £12.50

2. If one pound is worth $1.60, how many pounds do each of the following equal?

(a) $80   (b) $56   (c) $100   (d) $90

(Hint: to divide a number by 1.6, you can multiply it by 5 and then divide it by 8.)

3. If a South African rand is worth about 8p, how many rand is each of the following equal to?

(a) £15   (b) £90   (c) £1   (d) £30

4. What is one pound worth if:

(a) £40 is the same as 3,000 Indian rupees?

(b) £15 is the same as 165 Bolivian bolivianos?

(c) £300 is the same as 435 Swiss francs?

(d) £50 is the same as 450 Norwegian krone?

5. In each of these questions, I want to exchange £600. How much of the currency do I get?

(a) Argentine pesos at a rate of 7 pesos to the pound.
(b) Hong Kong dollars at a rate of $12.20 to the pound.
(c) Malaysian ringgit at a rate of 4.80 ringgit to the pound.
(d) Thai baht at a rate of 50 baht to the pound.

Mastering More Complicated Money Sums

Being able to do basic money sums is great – in real life, most of what you really need to do just involves simple arithmetic. However, in some situations you will need to work a bit harder to get the right answer.

Typical examples include dealing with percentages, working out prices from a complicated table, rounding off prices to get a rough estimate and – a great favourite of examiners – deposit schemes.

I cover working with percentages in detail in Chapter 8, and working with formulas and rounding in Chapter 4, so have a look at those chapters if you're uneasy about those. Meanwhile, I show you how to deal with a few examples involving comparing prices and deposit schemes.

Counting the cost

Dealing with price comparison questions can be a bit time-consuming, but if you set your work out neatly and take your time, they're not necessarily all that hard.

All you need to do is read the question carefully, figure out what you're 'buying' and work out the price in each of the offers.

For instance, if you had to buy a double bed from one of the two shops in Figure 10-5 and then get it delivered, you might say: 'At Sleepwell, the bed costs £650 plus another £50 for delivery – that makes £700. Then there's 10% off, so that makes £630 altogether. At Snore Draw, the bed costs £550 plus 10% delivery – so that's £615. So Snore Draw is £15 cheaper.'

```
┌─────────────────────────────┐ ┌─────────────────────────────┐
│ Sleepwell beds              │ │ Snore Draw furniture        │
│                             │ │                             │
│ King-size        £750.00    │ │ King-size        £650.00    │
│ Double           £650.00    │ │ Double           £550.00    │
│ Single           £500.00    │ │ Single           £400.00    │
│                             │ │                             │
│ Delivery         £50.00     │ │ Add 10% delivery charge     │
│                             │ │                             │
│ 10% off everything this     │ │                             │
│ weekend!                    │ │                             │
└──∧∧∧∧∧∧∧∧∧∧∧∧∧∧──┘ └──∧∧∧∧∧∧∧∧∧∧∧∧∧∧──┘
```

Figure 10-5: Looking at a typical price comparison table.

Dealing with deposit schemes

Some shops offer deposit schemes so that you can spread the payment for expensive goods over several months instead of having to pay everything up front at once. In real life, sadly, deposit schemes are usually quite complicated and it can be a struggle even for experts to figure out whether they're a good deal. Luckily, in the world of basic maths, things aren't quite so tricky.

A deposit scheme will usually be in two parts: a *deposit*, which is an amount of money paid up front, and *regular payments*, which are . . . well, the payments you make on a regular basis.

Here's what you need to know:

✔ The deposit may be given either directly, or as a percentage or a fraction of the full cost.

✔ The regular payments add up to the amount of each payment, multiplied by how many there are – so, 12 payments of £50 would make $12 \times 50 = £600$.

✔ The total amount you pay is the deposit plus all of the regular payments.

You may also need to work backwards. So, for example, to find the deposit if you're given the total and details of the payments, you simply take away all the payments from the total. Alternatively, to find the amount of the payments if you know the total and the deposit, you take the deposit away from the total and divide by the number of payments.

For example: suppose the total cost of a used car is £4,000. You pay a 25% deposit and then 10 equal payments to make up the balance – how much is each payment?

The deposit is £1,000, so the payments must come to £3,000. If there are 10 payments, they must each be 3,000 ÷ 10 = £300.

Attempting some more complicated money questions

You may need to link several methods together to answer the following sums. For example, if you have to pay a certain percentage deposit followed by a number of equal payments, you need to do the percentage sum first and then divide the end result by the number of payments.

1. It's sale time! How much would each of the following items be in the given sale?

(a) A calculator: normal price £10, 20% off sale
(b) A pair of walking boots, normally £80, 40% off
(c) Speakers, normally £45, 20% off
(d) A sofa, normally £750, 10% off
(e) A guitar, normally £200, 25% off
(f) A surfing lesson, normally £80, 15% off
(g) A bike, normally £300, 12% off
(h) a set of plates, normally £35, 20% off

Dummyjet		ColinAir	DAL	
Weekend away	£85pp	All holidays £25	Return travel:	£65 per person
One-week break	£155pp	per person per night		
Two-week break	£195pp		Hotel:	£30 per room per night
		No hidden fees!		
Luggage	£23 per bag		Luggage:	£15 per bag
			Admin fee:	£9.00

2. Looking at the offers in the above figure, you want to buy a one-week break for two people sharing a room with one suitcase.

(a) How much would the Dummyjet holiday package cost?

(b) How much would the ColinAir package cost?

(c) How much would the DAL package cost?

(d) Which offer is the best?

Newcastle	
Train ticket:	£103.90
Parking:	£4.30
Dinner:	£18.60
Hotel:	£65.75

Liverpool	
Petrol:	£75.83
Parking:	£23.50
Dinner:	£19.20
Hotel:	£83.20

3. Laura is checking her expenses from two trips; her receipts are shown in the above figure. She decides to check the amounts by rounding each figure to the nearest £10.

(a) After rounding, roughly how much did the Newcastle trip cost?

(b) After rounding, roughly how much did the Liverpool trip cost?

(c) Which trip was more expensive, and by roughly how much?

4. Jeremy is buying a new kitchen. If he pays up front, it costs £1,500. Alternatively, he can pay a 20% deposit and then 12 equal payments of £120.

(a) How big would his deposit be?

(b) How much would the equal payments come to?

(c) How much would he pay altogether under the deposit scheme?

(d) How much more expensive is the deposit scheme than paying up front?

5. Which formula would you use to check the price of a deposit scheme for a caravan for which the deposit was £1,000 followed by 10 equal payments of £800?

(a) $1000 + 800 \div 10$ (b) $1000 \times 10 + 800$

(c) $1000 + 800 \times 10$ (d) $(1000 + 80) \times 10$

Working through Review Questions

1. Work out:

(a) £2.25 + £64.69   (b) £22.30 + £98.62

(c) £81.34 − £46.23   (d) £17.90 − £13.66

(e) $3 \times$ £60.10   (f) $12 \times 98p$

(g) £86.44 $\div$ 4   (h) £69.99 $\div$ 3

2. Tickets for a museum cost £12.50 for an adult and £7.50 for a child. In total, how much would it cost for:

(a) Two adults?

(b) Three children?

(c) Two adults and three children?

(d) Three adults and a child?

3. The same museum introduces a family ticket, whereby two adults and two children can get in for £30.

(a) How much would it normally cost for two adults and two children?

(b) How much money would two adults and two children save as a result of getting the family ticket?

(c) What is the percentage reduction the family ticket offers?

(d) Would one adult and two children be better off paying separately or getting the family ticket?

4. Work out the following sales prices:

(a) A £15 fountain pen in a 10% off sale
(b) A £45 ticket to a football match with a 20% discount
(c) A £72 toolkit in a 25% off sale
(d) A £55 meal with a 10% off voucher

5. What percentage discounts are each of the following?

(a) A £400 computer on sale for £360
(b) A £250 bike on sale for £200
(c) A £140 suit on sale for £70
(d) A £50 desk on sale for £47.50

Cribbage's Pizza		Don Giovanni's	
Basic 9″ pizza:	£7.00	9″ pizza:	£9.00
Basic 12″ pizza:	£9.00	12″ pizza:	£12.00
Toppings:	50p each		
		All pizzas include two free toppings!	
Garlic bread	£3.20		
Chicken wings	£4.50	Garlic bread:	£3.00
		Chicken wings:	£4.00
All drinks:	£2.50		
		Bottled drinks:	£2.00
Delivery:	£3.00		
		10% discount if you collect!	
Special offer:			
Free delivery on orders over £20!			

6. The above figure shows two pizza delivery leaflets. Terry wants a 12″ mushroom and pineapple pizza, some garlic bread and a bottle of coke delivered.

(a) How much would his order cost from Cribbages?

(b) How much would his order cost from Don Giovanni's?

(c) How much more expensive is Cribbages?

(d) If Terry went to pick up the pizza, which restaurant would be cheaper, and by how much?

Waddle's Television sale!		Grimthorpe's Tellies	
26" TV	£170.00	26" TV	£220.00
32" TV	£190.00	32" TV	£240.00
36" TV	£220.00	36" TV	£265.00
(plus VAT at %20)			

7. The above figure shows the TV offers from two showrooms.

(a) How much would a 36" TV from Waddle's cost?

(b) How much would the same TV cost at Grimthorpe's?

(c) How much more expensive would the Grimthorpe's TV be?

(d) If you had a £50 gift voucher for Waddle's, how much cheaper would the Waddle's TV be?

Checking Your Answers

Check your answers and put any mistakes right as quickly as you can! Getting quick feedback about errors is one of the best ways to correct them.

Cash calculation questions

1. (a) £103.92   (b) £3.47   (c) £62.98   (d) £23.20
   (e) £161.23   (f) £51.24

2. (a) £13.20   (b) £17.40   (c) £142   (d) £6.83
   (e) £15   (f) £1.80

3. (a) $2 \times 7.50 = £15$ (b) $2 \times 3.90 = £7.80$
   (c) $2 \times 7.50 + 2 \times 3.90 = £22.80$
   (d) $3 \times 7.50 + £3.90 = £26.40$
   (e) $7.50 + 5 \times 3.90 = £27.00$

4. (a) $£18,000 \div 4 = £4,500$ (b) $£18,000 \div 12 = £1,500$
   (c) $£18,000 \div 10 = £1,800$ (d) $£18,000 \div 5 = £3,600$
   (e) $£18,000 \div 9 = £2,000$

5. (a) £16.01   (b) £24.45
   (c) $4 \times 1.20 = 4.80$, and $25 - 4.80 = £20.20$
   (d) Altogether, my groceries cost £14.34, so I expect
$25 - 14.34 = £10.66$ change

Currency questions

1. (a) €12   (b) €36   (c) €6   (d) €42
   (e) €144   (f) €66   (g) €76.80 (h) €15

2. (a) £50   (b) £35   (c) £62.50 (d) £56.25

3. (a) $1500 \div 8 = ZAR\ 187.50$ (b) $9000 \div 8 = ZAR\ 1125$
   (c) $100 \div 8 = ZAR\ 12.5$ (d) $3000 \div 8 = ZAR\ 375$

4. (a) £1 = 75 rupees   (b) £1 = 11 bolivianos
   (c) £1 = 1.45 francs   (d) £1 = 9 krone

5. (a) 4,200 pesos   (b) $7,320   (c) 2,880 ringgit
   (d) 30,000 baht

More complicated money questions

1. (a) $10 \times 80 \div 100 = £8$ (b) $80 \times 60 \div 100 = £48$
   (c) $£45 \times 80 \div 100 = 36$ (d) $£750 \times 90 \div 100 = £675$
   (e) $200 \times 75 \div 100 = £150$ (f) $80 \times 85 \div 100 = £68$
   (g) $£300 \times 88 \div 100 = £264$ (h) $35 \times 80 \div 100 = £28$

2. (a) A one-week break with Dummyjet costs £155 per person,
so for two people that's £310. The suitcase is an extra £23,
making £333.
   (b) With ColinAir, the cost is £25 each per night; so that's
£50 per night, times seven nights, making £350.
   (c) DAL will cost £65 each for the flight (£130) plus £30
times seven nights (£210) plus £15 for the luggage and £9 for
the admin. That's $£130 + £210 + £15 + £9 = £364$.
   (d) The best deal is Dummyjet.

3. (a) $£100 + £0 + £20 + £70 = £190$
   (b) $£80 + £20 + £20 + £80 = £200$
   (c) The Liverpool trip was about £10 more expensive (it was
actually £9.18 more expensive).

4. (a) £1,500 × 20 ÷ 100 = £300.   (b) £120 × 12 = £1,440.
(c) £300 + £1,440 = £1,740.   (d) £1,740 − £1,500 = £240

5. (C) You'd work out the total price as 1,000 + 800 × 10.

Review questions

1. (a) £66.94   (b) £120.92   (c) £35.11   (d) £4.24
   (e) £180.30   (f) £11.76   (g) £21.61   (h) £23.33

2. (a) £25   (b) £22.50   (c) £47.50   (d) £45

3. (a) £40   (b) £10   (c) 25%
   (d) One adult and two children would pay £12.50 + £15 =
   £27.50 normally, which is cheaper than the family ticket.

4. (a) £13.50   (b) £36   (c) £54   (d) £49.50

5. (a) 10%   (b) 20%   (c) 50%   (d) 5%

6. (a) The Cribbages pizza costs £9 plus 2 × 50p = £10. The
garlic bread is £3.20 and the coke is £2.50, making a total of
£15.70; so he'd have to pay the delivery charge of £3, making
£18.70.
   (b) Don Giovanni's pizza costs £12, plus £3 for the garlic
bread and £2 for the coke, making £17.
   (c) Cribbages is £1.70 more expensive.
   (d) Without the delivery charge, Cribbages costs £15.70;
with the collection discount, Don Giovanni's costs £15.30.
Don Giovanni's is 40p cheaper.

7. (a) £220 plus 20% VAT is £264.
   (b) £265, no funny business!
   (c) The Waddle's TV is £1 cheaper.
   (d) The Waddle's TV would be £51 cheaper.

Chapter 11

Working with Weights

● ●

In This Chapter

▶ Working out how to read scales

▶ Understanding different units and how to measure and compare them

▶ Using your knowledge of weights to make cooking and shopping easier

● ●

*W*orking with weight is one of the easiest bits of real-life maths. You don't have to deal with any funny business like you do with adding hours and minutes, and you don't have to worry about areas and volumes like you do with metres.

In this chapter, I show you how to read scales, do sums with weights, convert weights and figure out which of two offers is best.

Reading Scales

Reading a scale is very straightforward. Figure 11-1 shows a typical *analogue* scale (one with a dial; the ones with numbers that look like a digital watch are *digital* scales). Here's how you read it:

1. **Find the number to the left of the marker and the number to the right.** Here, that's 70 and 80.

2. **Count how many spaces are between the two marked numbers.** Here, there are 10.

3. **Divide the difference between the numbers in Step 1 by your answer from Step 2 to find out how much each tick is worth.** Here, it's one kilogram (kg).

4. **Count how many ticks higher or lower than the marked number the marker is.** Here, it's three – so the weight is 70 + 3 = 73kg.

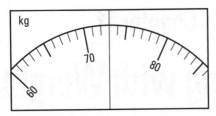

Figure 11-1: Reading an analogue scale.

Making sensible guesses

If the marker isn't exactly on a tick, you have to make a sensible guess about the weight. If you ignore the little ticks in Figure 11-1, you might say: 'The medium-sized tick is halfway between 70 and 80, so that must represent 75kg; the marker is slightly closer to midway than it is to 70, so 73 is a good guess.'

When you have to make a sensible guess like this, exams usually give you a little bit of leeway in your answer; so, if you said 72 or 74 for this question, you'd probably be okay.

Attempting some reading scales questions

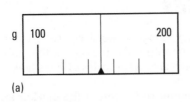

(a)

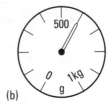

(b)

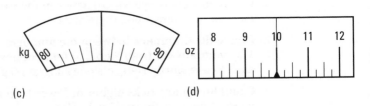

(c)

(d)

1. What weight is shown on scales (a), (b), (c) and (d) in the above figure?

2. An athlete keeps track of his weight. By how much did his weight increase or decrease between each pair of weighing-in sessions?

(a) Weighing 1: 85kg; weighing 2: 87kg.

(b) Weighing 1: 87kg; weighing 2: 84.7kg.

(c) Weighing 1: 84.7kg; weighing 2: 87.4kg.

(d) Weighing 1: 87.4kg; weighing 2: 87kg.

3. Work out the following:

(a) I buy 250g of coffee, but the sales assistant gives me an extra 50g to be nice. How much coffee do I get?

(b) I buy 500g of mushrooms and use 150g in a recipe. How many grams of mushrooms do I have left?

(c) I buy 4 cans of tomatoes, each weighing 440g. What weight of tomatoes do I have?

(d) I make a cake that weighs 1,200g, and split it into six equal slices. How much does each slice weigh?

Converting Weights

The only tricky thing about weights is that you can measure it in different units. This situation makes sense, if you think about it, because it keeps the numbers a reasonable size. You wouldn't want to have to measure your sugar in tonnes or an elephant in grams – you'd be working with either tiny or enormous numbers.

In this section, I introduce you to the three main types of metric weight (gram, kilogram and tonne), and the two main imperial weights (the ounce and the pound).

Technically, these metric units measure *mass* rather than weight, but unless you're an astronaut you really don't care.

Looking at different metric measures

The main metric unit you use for weight is the *kilogram* (or *kg*), which is about the same weight as a litre-bottle of water. Because kilo- means '1,000', one kilogram is the same as 1,000 grams (1,000g) – a gram is about the same weight as the sachets of sugar you get in a café. You can also use the tonne, which is 1,000 kilograms, or about the weight of a car. Here's how you switch between them:

✔ To convert from grams to kilograms or from kilograms to tonnes, you simply divide by 1,000.

✔ To convert from tonnes to kilograms or from kilograms to grams, you just multiply by 1,000.

When you convert to a bigger unit, you need fewer of them, so you divide; when you make the unit smaller, you need more of them, so you multiply.

Comparing metric and imperial measures

In the wizard currency described in the *Harry Potter* books one Galleon is worth 17 Sickles and one Sickle is worth 29 Knuts. According to Hagrid, the amiable half-giant, the system is 'easy enough'!

Personally, I think the author, J. K. Rowling, was having a bit of a pop at the imperial system, which also involves crazy numbers instead of the nice 10s, 100s and 1,000s of the metric system.

Imperial weights, for example, are measured in ounces (or *oz*) – an ounce is a bit less than 30 grams. A pound (or *lb*, a bit more than 450 grams) is 16 ounces; 14 pounds make a stone (*st*, about 6.4 kg). Eight stone make a hundredweight (*cwt*), and so on . . . all very simple, really.

Although the system sounds (and is) complicated, the maths you have to do with it isn't. You'll be given the conversion factors if you need to switch between metric and imperial units, at which point you can use the Table of Joy (see Chapter 7 for a full description of this handy tool). For instance, if you have to convert 600 grams into ounces (using the conversion one ounce = 30 grams), you need to follow these steps:

1. **Draw out your Table of Joy noughts and crosses grid, leaving plenty of room for labels.**

2. **Label the columns 'Grams' and 'Ounces', and the rows 'Conversion' and 'Weight'.**

3. **Fill in the numbers you know.** So, conversion/grams is 30, conversion/ounces is 1 and weight/grams is 600.

4. **Shade in the numbers like a chessboard and write down the Table of Joy sum.** So, $600 \times 1 \div 30$.

5. **Work out the sum.** The answer is 20.

Attempting some converting weights questions

1. What are the following weights in kilograms?

(a) 4,502 grams   (b) 56,215 grams   (c) 703 grams
(d) 0.5 tonnes

2. What are the following weights in grams?

(a) 1.5kg   (b) 0.35kg   (c) 15kg   (d) 0.001kg

3. What are the following weights in grams (one ounce is about 30 grams)?

(a) 4 ounces   (b) 10 ounces   (c) 16 ounces   (d) 1.5 ounces

4. What are the following weights in ounces?

(a) 240g   (b) 600g   (c) 90g   (d) 75g

5. What are the following weights in pounds (one pound is about 450 g)?

(a) 900g   (b) 4.5kg   (c) 135g   (d) 1,125g

6. What are the following weights in kilograms?

(a) 15 pounds   (b) 25 pounds   (c) 3 pounds   (d) 1.5 pounds

Shopping and Cooking using Weights

Most of the weight-related maths you may have to do in real life probably involves food and cookery. Quite often, you buy food by weight (looking in my cupboard, I've got a 500g pack of pasta, a 5kg bag of potatoes, a kilogram of sugar and so on); most recipes ask for a particular weight of certain ingredients.

In this section, I take you through two of the most common (and most involved) types of weight question: which of two offers is better value for money, and working out weight-related baking times.

Working out the price of eggs

Ever been in a supermarket and been flummoxed by all the yellow stickers saying 'three for the price of two' or 'buy one, get one free' or 'save 50p'? I could do my shopping in half the time if I didn't stop to work out the best deal – and I'd probably not have a house full of bananas that were on special offer. However, in exams, sometimes you *have* to figure such offers out – luckily, without the risk of cramming your house with cheap fruit.

Broadly speaking, you have two ways of figuring out which of two deals is better at your disposal. Both methods involve comparing like with like – you either want to compare the prices for the same *amount of stuff*, or compare the *amount of food* you get for the same amount of money.

For instance, the reason I have a house full of bananas is that my supermarket had 750g packages of bananas at three for the price of two – each package costing £1.20, while loose bananas cost £1 per kilo.

Which is the better value? First of all, I worked it out the 'same amount of stuff' way:

1. **Work out a way to buy the same amount of stuff.** Three bags of bananas weigh 2,250g, while the loose bananas are sold per kilo. If I bought 12 bags of bananas, that would be 9kg exactly.

2. **Work out the cost of each of the things.** If I want 12 bags, I only need to buy eight (and get four free), so that works out to 8 × £1.20 = £9.60. If I buy 9 kilos of loose bananas, that would cost 9 × £1 = £9.

3. **Decide which option is cheaper!** The loose bananas are the better value deal here.

Here's the other way to work out which is the better value: I figure out how I can spend the same amount of money:

1. **Work out how to spend the same amount of money.** In this case, I need to think in terms of a whole number of pounds. So, if three bags cost £2.40 (the price of two), I can buy 15 bags for £12.

2. **Work out how much stuff I get for my money.** So, 15 bags that weigh 750g each comes to 15 × 750 = 11,250g. For £12, I can also buy 12 kg of loose bananas.

3. **Decide which option gives me more stuff!** The loose bananas are still the better deal.

Other ways of working out similar sums are also available, but these two methods work well.

Calculating for cookery

A typical exam question involves cooking – for some reason – a turkey. It's always a turkey. Never a goose, a lasagne or a cake – always a turkey. And you'll be told that the recipe recommends cooking the turkey for a certain amount of time plus a certain other amount of time per kilogram. For instance:

> A recipe says to cook a turkey for one hour and 15 minutes, plus half an hour per kilogram. You have a 5.5 kg turkey; how long should you cook it for?

To answer that question, you have to follow these steps:

1. **Write down the time you have to cook the turkey for, no matter how big it is.** Here, that's 1:15.

2. **Work out how many extra minutes you need to cook it for.** You multiply the weight by the time per kilogram. That's $5.5 \times 30 = 165$ minutes, or 2:45.

3. **Add the two times together.** Here, you get 4 hours. (Chapter 9 covers time if you need some help.)

Attempting some shopping and cooking questions

1. In a supermarket, strawberries cost £4.80 per kilogram. How much would you have to pay for:

(a) 100g of strawberries?   (b) 400g of strawberries?
(c) 600g of strawberries?   (d) 250g of strawberries?

2. My favourite cheese is on sale for 90p per 100 g. How much would it cost if I bought:

(a) 200g?   (b) 400g?   (c) 250g?   (d) 50g?

3. The supermarket has several special offers on at any given time. For instance, pasta sauce is £3 for a 1 kg jar, or I can buy three 500g jars for £5.

(a) How much would 3kg of sauce cost if I bought the big jars?

(b) How much would 3kg of sauce cost if I bought the smaller jars?

(c) How many grams of sauce would I get for £15 if I bought the big jars?

(d) How many grams of sauce would I get for £15 if I bought the small jars?

(e) Which is better value?

4. A made-up recipe suggests I bake a turkey for one hour, plus 10 minutes per kilogram it weighs. How long should I bake the turkey if it weighs:

(a) 3kg? (b) 6kg? (c) 12kg? (d) 7.5kg?

I don't know a giblet from a goblet, so please don't follow that made-up recipe!

Working through Review Questions

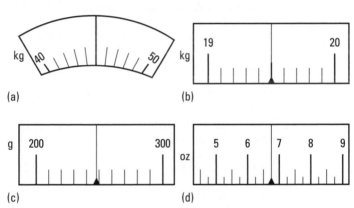

(a)   (b)

(c)   (d)

1. What weights are shown on scales (a), (b), (c) and (d) in the above figure?

2. Work out:

(a) 150g + 95g   (b) 45kg + 3.5kg   (c) 30kg – 2.35kg
(d) 350g – 96g   (e) 4 × 65g   (f) 16 × 200g
(g) 1,200g ÷ 4   (h) 1.2kg ÷ 4

3. (a) Carol's suitcase weighs 15kg and her backpack weighs 5.5kg. What is the total weight of her luggage?

(b) My plastic bags can hold a maximum of 3kg. How many 750g boxes of muesli can I fit in one bag?

(c) A runner used to weigh 90kg and then lost 7.3kg. How much did he weigh afterwards?

(d) I bought 250g of sweets and by the time I got home, there were only 175g left. What weight of sweets had mysteriously and inexplicably vanished?

(e) A recipe says to allow 75g of rice per person. How much rice do I need to feed six people?

(f) A lift can carry a maximum of 2,400kg. How many people can it hold, assuming they all weigh 80kg?

4. Convert the following into kilograms:

(a) 4,000g   (b) 3 tonnes   (c) 7,500g   (d) 750g

5. Convert the following into grams:

(a) 6kg   (b) 2.53kg   (c) 1.001kg   (d) 0.04kg

6. Convert the following into ounces, using the approximation one ounce = 30 grams.

(a) 420g   (b) 135g   (c) 90g   (d) 510g

7. Convert the following into grams:

(a) 4.5 ounces   (b) 20 ounces   (c) 9 ounces
(d) 10.5 ounces.

8. Convert the following into kilograms, using the approximation one pound = 450g.

(a) 4 pounds   (b) Three and a half pounds
(c) 200 pounds   (d) A third of a pound

9. Convert the following into pounds, using the approximation one kilogram = 2.2 pounds.

(a) 5kg   (b) 80kg   (c) 3.5kg   (d) one tonne

10. My local health food shop sells loose cous cous at £7.60 per kilogram. How much would it cost to buy:

(a) 100g?   (b) 500g?   (c) 250g?   (c) 750g?

11. I can buy loose cous cous at £7.50 per kilogram, or 300g packages for £2.50 each. Which is better value?

Checking Your Answers

The commonest mistake with weight questions is getting your units mixed up and trying to add kilograms to grams without converting first.

Reading scales questions

1. (a) 150g   (b) 600g   (c) 85kg   (d) 10 ounces

2. (a) An increase of 2kg   (b) a decrease of 2.3kg
   (c) an increase of 2.7kg   (d) a decrease of 0.4kg

3. (a) 250 + 50 = 300g   (b) 500 − 150 = 350g
   (c) 4 × 440 = 1760g (or 1.76kg)   (d) 1200 ÷ 6 = 200g

Converting weights questions

1. (a) 4.502kg (b) 56.215kg (c) 0.703kg   (d) 500kg

2. (a) 1,500g   (b) 350g   (c) 15,000g   (d) 1g

3. (a) 120g   (b) 300g   (c) 480g   (d) 45g

4. (a) 8 ounces   (b) 20 ounces   (c) 3 ounces   (d) 2.5 ounces

5. (a) 2 pounds

(b) 4.5kg = 4,500g; 4500 ÷ 450 =10 pounds

(c) 135 ÷ 450 = 27 ÷ 90 = 3 ÷ 10 = 0.3 pounds

(d) 1,125 ÷ 450 = 225 ÷ 90 = 45 ÷ 18 = 5 ÷ 2 = 2.5 pounds

6. (a) 15 × 450 = 4500 + 2250 = 6750g = 6.75kg

(b) 25 × 450 = 9000 + 2250 = 11,250g = 11.25kg

(c) 3 × 450 = 1,350g = 1.35kg

(d) 1.5 × 450 = 450 + 225 = 675g = 0.675kg

Shopping and cooking questions

1. (a) 48p   (b) 4.8 × 400 ÷ 1,000 = £1.92
   (c) 4.8 × 600 ÷ 1,000 = £2.88   (d) 4.8 × 250 ÷ 1,000 = £1.20

Once you've worked out 100g, you could also say '400g is 4 × 48p = £1.92' and so on; use whichever method you find easiest!

2. (a) 90p × 2 = £1.80   (b) 90p × 4 = £3.60
   (c) 90p × 2.5 = £2.25   (d) 90p × 0.5 = 45p

3. (a) Three 1kg jars would cost 3 × £3 = £9.

(b) Six 500g jars would cost £10.

(c) £15 would buy me 5 big jars, or 5kg of sauce.

(d) £15 would buy me nine small jars, or 4.5kg of sauce.

(e) Comparing the weights, 3kg of sauce costs less in the big jars; comparing the money, £15 buys me more sauce in the big jars. The big jars are better value.

4. (a) 1 hour and 30 minutes   (b) two hours   (c) three hours
   (d) two hours and 15 minutes

Review questions

1. (a) 45kg   (b) 19.5kg   (c) 247g   (d) 6.75 ounces

2. (a) 245g   (b) 48.5kg   (c) 27.65kg   (d) 254g
    (e) 260g   (f) 3,200g   (g) 300g   (h) 0.3kg

3. (a) $15 + 5.5 = 20.5$kg   (b) 3kg = 3000g, $3000 \div 750 = 4$
    (c) $90 - 7.3 = 82.7$ (d) $250 - 175 = 75$g
    (e) $75 \times 6 = 450$g   (f) $2,400 \div 80 = 30$ – a big lift!

4. (a) 4kg   (b) 3,000kg   (c) 7.5kg   (d) 0.75kg

5. (a) 6,000g   (b) 2,530g   (c) 1,001g   (d) 40g

6. (a) 14 ounces   (b) 4.5 ounces   (c) 3 ounces   (d) 17 ounces

7. (a) 135g   (b) 600g   (c) 270g   (d) 315g

8. (a) 1.8kg   (b) $3.5 \times 450 = 35 \times 45 = 1,575$g $= 1.575$kg
    (c) 90kg   (d) ⅓ of $450 = 150$

9. (a) 11 pounds   (b) 176 pounds   (c) 7.7 pounds
    (d) 1 tonne = 1,000kg, $1,000 \times 2.2 = 2,200$ pounds

10. (a) 76p   (b) £3.80   (c) £1.90   (d) £5.70

11. 300g of loose cous cous would cost 76p $\times 3 =$ £2.28, which is cheaper than the packet. The loose cous cous is better value, and more fun to say.

Chapter 12

Feeling the Heat: Getting to Grips with Temperature

In This Chapter

▶ Understanding how to read a thermometer

▶ Looking at three types of temperature question

▶ Considering Celsius and Fahrenheit

*I*f you listen to the radio or watch the TV, you almost certainly hear temperatures every day on the weather forecast. Most of the time – as long as you're not in the USA – forecasters give temperatures in *degrees Celsius* or °C; a temperature of 0°C means water will freeze or you'll see a frost; a warm summer's day is more like 20°C. Water boils at a temperature of 100°C.

In the olden days, and in America, you'd see temperatures given in *degrees Fahrenheit*, where water freezes at 32°F, a warm summer's day is about 70°F and water boils at 212°F, for reasons that are (bizarrely) all to do with Mrs Fahrenheit's armpit (see the 'Fiddling with Celsius and Fahrenheit' section for an explanation!).

In this chapter, I show you how to read a thermometer so that you can tell how warm it is; how to deal with temperature sums, including the dreaded negative numbers; and finally, how to convert between Celsius and Fahrenheit, in case you need to communicate with a time traveller or an American (or both).

Reading a Thermometer

A *thermometer* is a device you use to measure temperature. You get different types that measure different scenarios – the commonest ones measure the weather or the temperature of your body so you can tell whether you have a fever. These days, many thermometers are digital (meaning you can just read the digits on the display) rather than analogue (which involve reading a dial or scale). Most exam questions involve reading a scale, so here's how you do that:

1. **If more than one scale is presented, pick the correct one.** On a thermometer, you might see both Celsius and Fahrenheit – the question will tell you which you need to use if you have any doubt.

2. **If the dial or fluid level is next to a tick marked with a number, identify that number as your answer.**

3. **If the dial is next to a tick that's between two marked ticks, you need to figure out how much each tick is worth.**

4. **Divide the difference between the tick values by the number of *spaces* between them.** So, if four ticks appear between 30 and 40, that's five spaces – so each of them is worth 10 ÷ 5 = 2 degrees.

5. **Count how many spaces above or below the marked tick your dial is, and multiply this by your answer from Step 4.**

6. **If your value is higher than the marked tick, add your answer from Step 5 on to the marked tick; if it's lower, take it away.** That's your answer!

Attempting some reading a thermometer questions

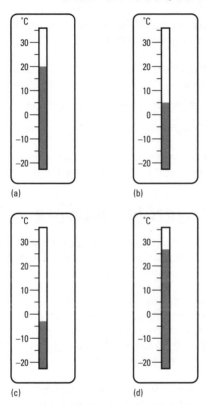

1. What temperatures are shown on thermometers (a), (b), (c) and (d) in the above figure?

2. If it's a balmy 17°C in Dorset, what temperature is it:

(a) In Dundee, where it's 15°C cooler?

(b) In Barcelona, where it's 5°C warmer?

(c) In London, where it's 3°C warmer?

(d) In Reykjavik, where it's 8° cooler?

Looking at Temperature in Three Different Ways

The range of sums you can do with temperatures is actually quite limited. You may be asked to:

✔ Find the temperature after (or before) it's changed by a certain amount.

✔ Find the difference between two temperatures.

✔ Put a list of temperatures into order.

The only slightly tricky thing is working with negative numbers – and I take you through that carefully here.

Nailing negative numbers

Getting your head around negative numbers can be challenging, but once you can put them into context – especially with temperature – they become much easier.

If you live anywhere in Britain for more than a year, I guarantee you'll experience negative temperatures – when it gets colder than 0°C. One degree colder than freezing is –1°C, two degrees colder than freezing is –2°C, 12 degrees colder than freezing is –12°C . . . you get the picture.

Figure 12-1 shows a number line including negative numbers, which you'll be using shortly to do some sums.

$$-10 \quad -9 \quad -8 \quad -7 \quad -6 \quad -5 \quad -4 \quad -3 \quad -2 \quad -1 \quad 0 \quad 1 \quad 2 \quad 3 \quad 4 \quad 5 \quad 6 \quad 7 \quad 8 \quad 9 \quad 10$$

Figure 12-1: Looking at a number line.

You need to think about sums on the number line in this way:

✔ If you're adding something on, you move to the right.

✔ If you're taking something away, you move to the left.

✔ *But* if the thing you're adding on is negative, you must do the reverse of what I just said.

So, to work out 6 – 8, you start at 6 and move eight spaces left, ending up with –2. For –6 – 8, you would start at –6 and move eight spaces to the left, ending up at –14. To figure out –6 + 8, you start on –6 and move eight to the *right*, ending up on 2.

To avoid any more confusion than is completely necessary, I've put brackets around negative numbers in sums so you don't see monstrosities like –3 – –6, which is really hard to read.

How about 4 + (–6)? Well, you start on 4 as normal, think 'it's a plus so I'll move right . . . but wait! It's a negative number, so I have to move the other way.' So you move *left* six and end up on –2. To sort out (–7) – (–3), you start on –7 and end up moving *right* three and get to –4.

Warming up and cooling down

Working out the temperature after an increase or decrease is a walk in the park! If you apply a little bit of common sense, the answers to these types of sum become really obvious. A typical question could ask:

The temperature last night was 3°C. This morning, it is 7°C warmer; what temperature is it now?

Here's the recipe for answering a question like that:

1. **Work out whether the temperature you want is warmer or colder than the temperature you're given.** Here, the temperature now is warmer than the 3°C you're given.

2. **If the temperature you want is warmer than the one you're given, add the temperature change on to the initial temperature.** Here, you work out 3 + 7 = 10°C to get your answer.

3. **If the temperature you want is cooler than the one you're given, take the temperature change away from the temperature you know.** That number is your answer.

Sometimes you may start or end in negative numbers. Don't worry; just draw out a number line if you need to.

Working out differences between temperatures

Finding the difference between two temperatures is easier than working out increases and decreases in temperature, if you ask me – you don't even need to worry about what kind of sum it is (it's always a take away).

You may see a question looking something like this:

> The temperature in Lerwick is 9°C. In Cardiff, it's 14°C. How much warmer is Cardiff than Lerwick?

You don't even need to use a recipe to work out the answer! All you do is take away the lower temperature from the higher one. Cardiff is 14 – 9 = 5°C warmer than Lerwick.

Putting temperatures in order

You may also be asked to put some temperatures in order. This is the kind of task that only ever comes up in exams; I can't think of any situation in real life in which you'd need to put temperatures in order. Other things, certainly, but temperatures? If you can think of a good reason to do so, I'd love to hear about it!

Here's how to deal with such a question: let's say you're asked (unreasonably) to put the temperatures –2°C, 2°C, 5°C, –7°C and 9°C in order:

1. **Find the coldest temperature in the list and write it down. Now circle the number in the original list.**
   Here, it's –7°C.

2. **Find the coldest temperature in the list that you haven't circled yet. Write it to the right of your previous answer, and circle it in the original list.**

3. **Repeat Step 2 until you've circled all of the numbers.**
   Your sorted list should look like this: –7°C, –2°C, 2°C, 5°C and 9°C.

Be careful of negative temperatures, as always – any negative temperature is colder than any positive temperature, no matter what the numbers are!

Attempting some temperature questions

1. What is the temperature now, if:

(a) It was 6°C and cooled by 8°C?

(b) It was 4°C and cooled by 10°C?

(c) It was –2°C and warmed up by 5°C?

(d) It was –5°C and warmed up by 2°C?

(e) It was 10°C and cooled down by 10°C?

(f) It was –10°C and warmed up by 10°C?

2. What is the temperature difference between:

(a) My freezer, where it's –10°C, and my kitchen, where it's 20°C?

(b) Oslo, where it's –4°C, and Nice, where it's 12°C?

(c) Helsinki, where it's –9°C, and Moscow, where it's –12°C?

(d) Glasgow, where it's 0°C, and Toronto, where it's –15°C?

3. Put the following lists of temperatures in order:

(a) 1°C, –2°C, 3°C, –4°C, 5°C

(b) 7°C, –7°C, 0°C, 3°C, –3°C

(c) 12°C, –5°C, –7°C, 5°C, –3°C

(d) –4.3°C, –4.7°C, 4.2°C, 4.8°C, 0.4°C

Fiddling with Celsius and Fahrenheit

Back in the eighteenth century, a universally-accepted scale for measuring temperature didn't exist. Before that, people just said 'it's a bit chilly' or 'what a nice day'. Shakespeare didn't write 'Shall I compare thee to a summer's day? Thou art 37°C and bring out my hay-fever.'

Luckily, along came two scientists – Fahrenheit (from modern-day Poland) and Celsius (from Sweden) – who each came up with ways of measuring temperature. Unfortunately, they used different scales: Celsius decided that water should freeze at 100° and boil at 0° – the other way round from today; Fahrenheit made 0° the coldest temperature he could concoct with ice and salt, and 100° was the temperature of his wife's armpit.

So, clearly a few issues cropped up with these early methods for dealing with temperature. Over time, they developed into something a little more sensible:

✔ In Celsius, water freezes at 0°C and boils at 100°C. A healthy human body has a temperature of about 37°C.

✔ In Fahrenheit, water freezes at 32°F and boils at 212°C. A healthy human body has a temperature of about 99°F.

Some scientists also use a scale named after Kelvin, but you *really* don't care about the Kelvin scale, where water freezes at 273.13K and boils at 373.13K. It's quite useful for extremely cold temperatures, though!

The degree Celsius is used across most of the world – except in America, which clings stubbornly to the degree Fahrenheit. Until about 20 years ago, it was in common currency in the UK, too; you sometimes still hear Fahrenheit temperatures used by people of my parents' generation.

Luckily, though, you don't see them much in numeracy exams except as an excuse to practise conversion! You may need to use a chart, a table or a formula to convert from one kind of temperature to the other.

Temperature is the one kind of conversion where the Table of Joy *doesn't* apply at all. This is all Fahrenheit's fault – if he'd started his scale with water freezing at zero, you'd have been in the clear!

Converting from a graph or table

Converting Fahrenheit to and from Celsius using a graph or a table is straightforward. I give you a brief outline here, but you can check out Chapter 15 on graphs and tables if you want more explicit instructions! Figure 12-2 shows the two methods for converting Celsius and Fahrenheit and I describe them below.

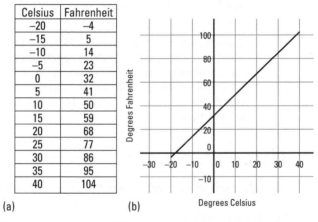

Celsius	Fahrenheit
−20	−4
−15	5
−10	14
−5	23
0	32
5	41
10	50
15	59
20	68
25	77
30	86
35	95
40	104

(a)   (b)

Figure 12-2: Converting Fahrenheit and Celsius using (a) a graph and (b) a table.

Here's how you use a graph to convert:

1. **Find the axis representing the temperature you're given.**

2. **Find the temperature you're given on that axis.**

3. **Draw or imagine a straight line from the temperature you just found, going directly up or straight to the right until it hits the line of the graph.**

4. **Draw or imagine a line at right angles to your previous line, going directly to the other axis.**

5. **Read the value off where this line meets the other axis – this is your answer!**

With graph questions, you're usually given a little bit of leeway; if you're off by one or two degrees, it shouldn't be a problem!

Using a table is just as easy:

1. **Look at the column that contains the temperature you're given.**

2. **If the number you're given is in the table, you just need to read the number beside it.** That is your answer.

3. **If the number you're given isn't in the table, you need to *interpolate* – make a sensible guess based on the nearby numbers.** For instance, if your number is midway between two numbers in the table, you should give a number midway between the numbers next to them.

Using a formula

The other way to convert between Celsius and Fahrenheit is to use a formula. Several ways of writing the formulas exist, but the commonest are:

✔ To find the Celsius temperature: $C = (F - 32) \times 5 \div 9$.

✔ To find the Fahrenheit temperature: $F = C \times 9 \div 5 + 32$.

You don't have to remember these formulas; if you need them, they'll be given to you.

Chapter 4 provides a more detailed explanation of formulas, but here are the steps involved in creating a formula to work out a temperature conversion:

1. **Pick the formula that gives you the kind of degrees you're looking for.**

2. **If you're given the Fahrenheit temperature, replace the letter *F* with the number of Fahrenheit degrees.**

3. **If you're given the Celsius temperature, replace the letter *C* with the number of Celsius degrees.**

4. **Work out the sum after the equals sign.** You need to follow the BIDMAS rule described in Chapter 4 to make sure you work out the elements of the sum in the correct order. Your result is the answer!

Attempting some Celsius and Fahrenheit questions

1. Use the graph in Figure 12-2(a) to estimate:

(a) 3°C in Fahrenheit   (b) 9°C in Fahrenheit

(c) 44°F in Celsius   (d) 93°F in Celsius

2. Use the table in Figure 12-2(b) to work out:

(a) –40°F in Celsius   (b) 32°F in Celsius

(c) 50°F in Celsius   (d) 86°F in Celsius

(e) 30°C in Fahrenheit   (f) 10°C in Fahrenheit

(g) 25°C in Fahrenheit   (h) –20°C in Fahrenheit

3. Use the formula $C = (F - 32) \times 5 \div 9$ to work out the Celsius temperature equivalent to:

(a) 77°F   (b) 95°F   (c) 41°F   (d) 185°F

4. Use the formula $F = C \times 9 \div 5 + 32$ to work out the Fahrenheit temperature equivalent to:

(a) 200°C   (b) 60°C   (c) 15°C   (d) 95°C

Working through Review Questions

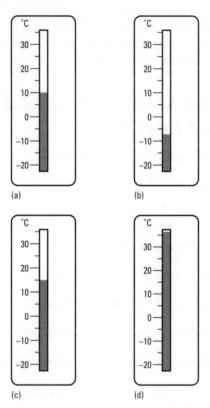

(a)

(b)

(c)

(d)

1. What temperature is shown on thermometers (a), (b), (c) and (d) in the above figure?

2. What is the difference between:

(a) 190°C and 220°C?   (b) –15°C and –12°C?

(c) –15°C and 12°C?   (d) –12°C and 15°C?

3. What is the temperature now, if:

(a) It was 7°C and it dropped by 5°C?

(b) It was –7°C and it dropped by 5°C?

(c) It was –7°C and it rose by 5°C?

(d) It was 7°C and it rose by 5°C?

4. Use the formula $F = C \times 9 \div 5 + 32$ to find the Fahrenheit temperature that's the same as:

(a) 45°C   (b) 55°C   (c) 100°C   (d) 35°C

5. Use the formula $C = (F - 32) \times 5 \div 9$ to find the Celsius temperature that's the same as:

(a) 68°F   (b) 95°F   (c) 113°F   (d) 41°F

6. Put the following temperatures in order:

(a) 10°C, –4°C, 7°C, –6°C, 12°C

(b) 5°C, –1°C, –10°C, 12°C, 0°C

(c) 9°C, –8°C, 7°C, –5°C, 3°C

(d) –4°C, –3.5°C, –4.5°C, –5°C, 4°C

Checking Your Answers

Be careful not to get caught out with the negative numbers questions – draw out a number line if you're not sure of the answer!

Reading a thermometer questions

1. (a) 20°C   (b) 5°C   (c) –3°C   (d) 27°C

2. (a) 2°C   (b) 22°C   (c) 20°C   (d) 9°C

Temperature questions

1. (a) –2°C   (b) –6°C   (c) 3°C   (d) –3°C
   (e) 0°C   (f) 0°C

2. (a) The kitchen is 30°C warmer
   (b) Nice is 16°C warmer (which is Nice)
   (c) Helsinki is 3°C warmer
   (d) Glasgow is 15°C warmer

3. (a) –4°C, –2°C, 1°C, 3°C, 5°C

(b) –7°C, –3°C, 0°C, 3°C, 7°C

(c) –7°C, –5°C, –3°C, 5°C, 12°C

(d) –4.7°C, –4.3°C, 0.4°C, 4.2°C, 4.8°C

Celsius and Fahrenheit questions

1. (a) 37°F   (b) 48°F   (c) 7°C   (d) 34°C

(Within one or two degrees is acceptable.)

2. (a) –40°C (b) 0°C   (c) 10°C   (d) 30°C
   (e) 86°F   (f) 50°F   (g) 77°F   (h) –4°F

3. (a) 25°C   (b) 35°C   (c) 5°C   (d) 85°C

4. (a) 392   (b) 140°F   (c) 59°F   (d) 202°F

Review questions

1. (a) 10°C   (b) –7°C   (c) 15°C   (d) 37°C

2. (a) 30°C   (b) 3°C   (c) 27°C   (d) 27°C

3. (a) 2°C   (b) –12°C   (c) –2°C   (d) 12°C

4. (a) 113°F   (b) 131°F   (c) 212°F   (d) 95°F

5. (a) 20°C   (b) 35°C   (c) 45°C   (d) 5°C

6. (a) –6°C, –4°C, 7°C, 10°C, 12°C

(b) –10°C, –1°C, 0°C, 5°C, 12°C

(c) –8°C, –5°C, 3°C, 7°C, 9°C

(d) –5°C, –4.5°C, –4°C, –3.5°C, 4°C

Chapter 13

Sizing Up Shapes

· ·

· ·

*T*his chapter is all about how *big* things are. 'Bigness' is a pretty vague and flexible concept – for instance, if you saw a man who was six-foot-three (190cm) tall walking down the street, you might think 'that's a big lad!', but if you saw him in a line-up of professional basketball players, you might think 'what a titch!'

So instead of simply describing people – and things – as 'big' or 'small', it's much better for communicating if you put a number on them (like I did when I said the man was 190cm tall).

Even then, you can measure three different kinds of bigness:

✔ The *length* of a thing, which is how many rulers of a given size you can place alongside it.

✔ The *area* of a thing, which is how many tiles of a given size you would need to cover it.

✔ The *volume* or *capacity* of a thing, which is how many dice of a given size you would need to fill it.

Volume and capacity aren't quite the same thing, strictly speaking, but you'd only care about the difference in a very restricted set of circumstances. Volume is really how much space something takes up and capacity is how much fits inside it – as far as you need to care, they're the same thing.

Working in Different Units of Length

I talk a little bit about the metric system in the other chapters in this part of the book – apart from time, nearly everything you measure in modern maths uses units based on tens, hundreds and thousands.

Length is no different. The basic measure of length is the *metre*, usually abbreviated to *m*, which is about the distance between the floor and the handle on a normal-sized door.

That's not a practical measurement for things much smaller than a door, so you can split up a metre into 100 *centimetres*, or *cm*, each of which is about the width of a finger – they're also the big ticks on a normal ruler. You can also split a metre into 1,000 *millimetres*, or *mm*, which are the smaller ticks on a ruler.

Going the other way, if you had to talk about the distance between towns in metres, you'd soon run into huge numbers; instead, you'd typically use *kilometres*, or *km*, each of which is 1,000 metres – two and a half laps of an athletics track.

You may also need to know about some more old-fashioned units of length:

✔ An *inch* is about 2.5 centimetres, which is the size of the 'tab' key on my keyboard.

✔ A *foot* is 12 inches, or a little over 30 centimetres. A sheet of A4 paper is about a foot long.

✔ A *yard* is 3 feet, or a little less than a metre.

✔ A *mile* (probably the most common of these still in everyday use) is 1,760 yards, or about 1.6 kilometres.

Almost all of the questions you'll encounter will involve centimetres and metres. You may get questions involving miles, and possibly miles per hour, but the odds are you'll only see the others if you're converting them into the metric system. See the section on 'Converting lengths', later in this chapter, to learn how to do that!

Attempting some unit of length questions

1. Pick the correct *metric* unit of length for the following:

(a) An athletics track is 400 __ all the way round.

(b) A door is about 200 __ high.

(c) London and Edinburgh are about 650 __ apart.

(d) A ruler is about 30 __ long.

2. Pick the right *imperial* unit of length for the following:

(a) This book is about 11 __ long.

(b) A door is about 7 __ high.

(c) London and Birmingham are about 100 __ apart.

(d) A football pitch is about 100 __ long.

Looking out for length

A silly old riddle says, 'how long is a piece of string?', to which the answer is 'twice as long as half a piece of string'. Apart from the practice with fractions, it's pretty much useless as an answer. Putting a number on that piece of string is a better idea.

You can measure the length of an object – if it's straight, at least – by putting a ruler (or tape measure, or something similar) alongside it and counting the tick marks on the ruler. Alternatively, if your object isn't straight, you can get a bit of string (or anything else that's long and flexible), run it along the thing you want to measure, mark the string at the start and end of the object and then straighten out the string and measure it.

Keeping your distance

There's not much difference between the idea of a length and the idea of a distance: a length usually describes the size of an object, while a *distance* usually describes how far apart two objects are. A distance is generally between things, rather than along things.

Distance uses the same units as length – you'd normally measure a distance in metres, or kilometres (maybe miles) if it's a long way, or centimetres if it's shorter. You should always describe lengths and distances in the most practical unit!

Walking around the perimeter

You may have heard the phrase *perimeter fence*, which just means a fence that goes all the way around something – usually a military base or an airport. Unsurprisingly, that's roughly what *perimeter* means, too – it's the distance all the way around an object. Finding out the perimeter of a shape with straight sides (like the one in Figure 13-1) is easy. Just follow these steps:

1. **Work out the length of each of the sides of the shape.** The lengths of the sides of the shape in Figure 13-1 are 7cm, 12cm, 3cm, 4cm, 4cm and 8cm.

2. **Add them up.** The shape in Figure 13-1 has a perimeter of 38cm.

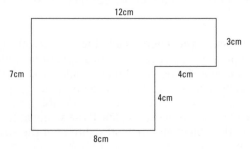

Figure 13-1: Finding the perimeter of a shape.

That's it! And here's the good news: the numeracy curriculum doesn't cover shapes with curvy sides, so you now know everything you need to know about perimeters.

I like to mark off the sides on the picture as I add them up, just so that I don't miss any or count any of them twice.

Attempting some length, distance and perimeter questions

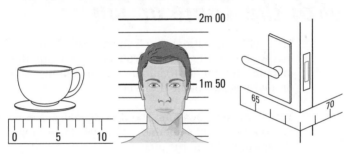

1. Look at the above figure, which shows several objects next to measuring devices. (The ruler and the tape-measure are marked in centimetres.)

(a) How wide is the coffee cup?

(b) How tall is the man?

(c) How wide is the door?

2. A roll of tin foil is 20m (2000cm) long. On average, I use 25 centimetres of tin foil at a time. How many uses should I expect to get out of the roll of foil?

3. A runner is training for a marathon. On Monday, she runs 20km. On Tuesday, she runs 5km. On Wednesday and Thursday, she runs 15km each day. On Friday, she covers 10km before the 40km race on Saturday. How far did she run altogether in the week?

4. A golf course has 18 holes, with an average length of 400 metres each. How far would the golfer have to walk to cover the whole course?

5. A village green is a rectangle. The short sides are 70m long and the long sides are 120m long. What is the perimeter of the village green?

Working out Length Conversions with the Table of Joy

Many of the questions you'll come across in numeracy tests involve converting lengths from one unit to another – for instance, from miles to kilometres, or centimetres to inches, or anything else along those lines.

You'll also see questions about using map scales – either to find out how long something is in real life, or how long something should be on the map.

You may need both of these skills in real life once in a while. I was out walking the other day and had to read the map to find out how far we were from the car, and then translate that distance into miles for my old-fashioned friend.

You can work out both types of scale problem using the Table of Joy, which I introduce in Chapter 7. Have a brush up on that chapter if you need to but I show you another way of doing these sums, too.

Converting lengths

To work out the answer to a conversion sum, you need to know the *conversion factor* in the question, that is, how many of one thing make the other thing (for instance, 91.4 centimetres make one yard). Here are some conversions that I'd recommend learning:

✔ 1 kilometre is 1,000 metres, or 100,000 centimetres, or 1,000,000 millimetres.

✔ 1 metre is 100 centimetres, or 1,000 millimetres.

✔ 1 centimetre is 10 millimetres.

✔ 1 inch is about 2.5 centimetres.

✔ 1 foot (12 inches) is about 30 centimetres.

✔ 1 mile is about 1.6 kilometres, or 1,600 metres.

You *may* be given the last three of those conversions if you need them to answer an exam question, but I'd learn them all the same!

You may have noticed that the first three conversions in that list all involve tens, hundreds and so on – all very nice numbers to deal with! The last three . . . not so much. You can make conversion sums easy in two different ways: a simple, direct way when you're changing between metric units (like the first three), and a bit of the Table of Joy for the slightly tougher last three.

Keeping it simple with the metric system

The easiest way to convert between different metric units is simply to multiply or divide by tens until you get the correct unit. I cover multiplying and dividing by powers of 10 in Chapter 3.

Let's say you want to convert 3,600 metres into kilometres. Follow these steps:

1. **Decide whether the units you want are bigger or smaller than the units you're given.** Here, you want kilometres, which are bigger than metres.

2. **Work out how many times bigger or smaller the units are.** You know, because you memorised the bullets, that one kilometre is 1,000 metres.

3. **If the units you want are *bigger*, *divide* by your answer from Step 2. Otherwise, *multiply* by the answer from Step 2.** Here, you work out 3,600 ÷ 1,000 = 3.6.

If you don't like that method, use the Table of Joy approach in the next section.

Taking on imperial units with the Table of Joy

To convert between two different measurement units with the Table of Joy, for example, to work out how many kilometres a seven-mile race is, follow these steps:

1. **Draw out a noughts and crosses grid and label the columns with the names of the units ('Miles' and 'Kilometres'). Then label the rows 'Conversion rate' and 'Distance'.**

2. **Fill in the numbers you know.** One mile is 1.6 kilometres, so write 1 in the conversion rate/miles box and 1.6 in the conversion rate/kilometres box. You want to convert 7 miles, so put 7 in the distance/kilometres box.

3. **Shade in the grid like a chessboard and write down the Table of Joy sum.** Multiply the numbers on the same shaded squares and divide by the number in the other square. Here, you get $7 \times 1.6 \div 1$.

4. **Work out the answer!** Seven miles is the same as 11.2 kilometres.

You can also convert backwards using the same idea. At the end of Step 2, you just put the thing you want to convert in the correct box and the Table of Joy takes care of the rest.

Keeping to scale

A very common question in adult numeracy tests involves translating a distance on a map with a certain *scale* into a distance in real life. When you draw an accurate map, you have to reduce everything in size (a map of Britain that was the size of Britain wouldn't be terribly practical, right?) and normally one centimetre on the map represents a certain number of kilometres in real life. (You also see some maps as 'inches to the mile', which is the same idea.) The scale is simply how much you multiply the map distance by to get the real-life distance.

For instance, an exam question may ask something like:

A map has a scale of 1:50,000. A path on the map has a length of 3cm. How long is the path in real life (in kilometres)?

You can work out this conversion using the Table of Joy, too. Follow these steps:

1. **Draw out the noughts and crosses grid and label the columns as 'Map' and 'Real life'. Label the rows as 'Scale' and 'Path' (or whatever you're measuring).**

2. **Fill in the numbers you know.** So, 1cm on the map is 50,000cm in real life, so put 1 in the map/scale square and 50,000 in the real life/scale square. The path is 3cm on the map, so put 3 in the map/path square.

3. **Shade in the table like a chessboard and write down the Table of Joy sum.** Multiply the two numbers on the same shaded squares and divide by number in the other square. That's $3 \times 50,000 \div 1$.

4. **Work out the sum.** Here, it's 150,000cm.

5. **Convert this answer into the units the question wants.** So, 150,000cm is the same as 1,500m or 1.5km.

Converting from real-life distances to map distances is just as easy. You simply put the real-life distance in the relevant square at the end of Step 2.

Attempting some conversion and scale questions

1. Convert the following:

(a) 0.35 metres into centimetres.

(b) 100 kilometres into metres.

(c) 40 millimetres into centimetres.

(d) 15,000 millimetres into metres.

2. Work out:

(a) If one inch is 2.5 centimetres, how far is 16 inches in centimetres?

(b) How far is 100 centimetres in inches?

(c) If one mile is 1,600 metres, how far is 8 kilometres in miles?

(d) How far is 8 miles in metres?

3. Finally, some questions about scales:

(a) A map has a scale of 1:50,000. A path on the map is 4cm long. How long is the path in real life?

(b) A different map has a scale of 1:10,000. If I walk 16 kilometres, how far would that be on the map?

(c) A map of the UK has a scale of 1:1,600,000. Two cities are 160 km apart in real life. How far apart are they on the map?

Filling Things Up

Try this brief experiment. Find two glasses – one wide and one narrow. Fill the wide one up to a depth of one centimetre (or one finger width if you don't have a ruler handy). Pour the water into the narrow glass. How many finger widths high is the water now?

That little experiment is why you need a different kind of measurement when you're thinking about *volume* (how much space does something take up?) instead of length. You used the same volume of water in both glasses (you didn't spill any, right?) but the depths were completely different.

You measure volume in units of millilitres (*ml*), which are also called *centimetres cubed* (cm^3). A *litre* – about three cans of fizzy pop – is 1,000 millilitres.

You also need to know about a couple of old-fashioned imperial units: the *pint*, which is 568 millilitres – a bit more than half a litre. In the UK, you normally buy beer in pints (sometimes half-pints or thirds of a pint) but almost everything else is sold in millilitres. Some traditions die hard!

You may also come across the gallon, which is eight pints or about 4.5 litres. You don't see gallons very often these days, because litres are much easier to work with. (Dividing by 10 is easier than dividing by 8, don't you think?)

If you're asked about pints or gallons, you'll be given the relevant conversion rates. Even I had to look them up!

Working out the volume

To work out how many millilitres (or cubic centimetres) a box holds, you can use a very simple recipe:

1. **Write down the height, width and length of the box.**
2. **Multiply them together.**

So, if you had a box that was 4cm long, 2cm high and 3cm wide, you would work out $4 \times 2 \times 3 = 24cm^3$. The answer is the volume.

Rarely do you see a more complicated volume question than a simple box at the level of basic maths. The only things you may come across are compound boxes – two boxes of different sizes stuck together – or possibly working with a given formula. Compound boxes are very straightforward: you just split the box up into its component parts, find the volume of each and then add them up. Chapter 4 provides the lowdown on working out sums using formulas.

Battling with boxes

Examiners are unusually fond of one other kind of box-related question, too: it involves working out how many things of a particular size fit in a box. Honestly, they're obsessed with efficient packing. A typical question may ask:

A box has a width of 100cm, a length of 50cm and a height of 40cm. Alex wants to stack this with books that are 30cm wide, 25cm long and 4cm tall. How many books can he fit in the box?

And here's the recipe for working out the answer to this kind of question:

1. **Divide the length of the box by the length of the object going inside it. Write down the number, but ignore any remainder.** Here, you do 100cm ÷ 30cm = 3 ⅓, so you write down 3.

2. **Divide the width of the box by the width of the object going inside it. Write down the number, but ignore any remainder.** Here, you do 50cm ÷ 25cm = 2, so you write down 2.

3. **Divide the height of the box by the height of the object going inside it. Write down the number, but ignore any remainder.** Here, you do 40cm ÷ 4cm = 10, so you write down 10.

4. **Multiply your three answers together.** You get 3 × 2 × 10 = 6 × 10 = 60 books that Alex can fit in his box. Never mind whether he can lift it afterwards!

Attempting some volume and capacity questions

1. What is the volume of a box with the following dimensions?

(a) 30cm high, 40cm wide, 50cm long.

(b) 20cm wide, 5cm tall, 30cm long.

(c) 10cm wide, 10cm tall, 10cm long.

2. The figure below shows a building made of two cuboids joined together. The bigger one is 4m tall, 5m wide and 10m long. The smaller one is 2m tall, 3m wide and 5m long. The building's owner needs to work out its volume to estimate her heating bill. What is the building's volume?

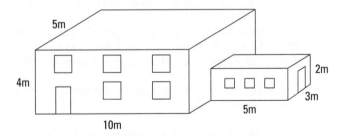

3. A box is 200cm long, 50cm tall and 75cm wide.

(a) What is the volume of the box?

(b) This book is about 20cm long, 5cm tall and 15cm wide. How many copies of the book would fit in the box?

(c) A Rubik's cube is 10cm wide, 10cm tall and 10cm long. How many Rubik's cubes could fit in the box?

Calculating Area

You can think of *area* as the amount of wrapping paper you'd need to cover a shape – assuming you wrap things like I do, by cutting out rectangles just big enough to cover the whole shape and then joining the ends of the paper together with sticky tape.

You normally measure area in centimetres squared (cm^2). One centimetre squared, like the name suggests, is the area of a square that's one centimetre long on each side. You may also see areas in metres squared (or m^2 – one metre squared is the same as the area of a square that's one metre long on each side), or kilometres squared (km^2). Whenever you're working out an area, you should make sure the units are the same on both sides, and give your answer in that unit squared.

You only need to know the area of one shape for basic maths, and it's an easy one: to find the area of a rectangle, you multiply the width by the height. Looking at the rectangles in Figure 13-2, the one on the left has an area of $15cm^2$ ($3cm \times 5cm$) and the one on the right has an area of $91cm^2$ ($7cm \times 13cm$).

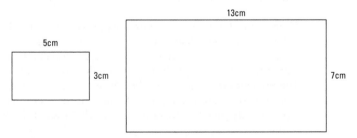

Figure 13-2: Working out the area of two rectangles.

Knowing the area of a triangle is also quite nice because it's almost as simple: all you do is multiply the width by the height and divide your answer by two. (That's because, if you cut a rectangle in half, corner to corner, you get a triangle. There's a bit more to it than that, but that's the basic idea.)

Attempting some area questions

1. What is the area of a rectangle with the following dimensions?

(a) 12cm long and 5cm wide?

(b) 6cm long and 2.5cm wide?

(c) 10cm long and 12cm wide?

2. A football pitch is 100 metres long and 70 metres wide. What is its area in square metres?

3. The nice, rectangular state of Colorado is about 600 kilometres wide and 450 kilometres long. What is the area of Colorado, in square kilometres?

Compounding shapes

Sadly, the basic maths examiners very rarely want you to do something as straightforward as working out the area of a rectangle. Doing so would be just too easy. Instead, they tend to throw out *compound shapes* for you to figure out, which basically means two or three rectangles stuck together, like the shapes in Figure 13-3.

Working out a compound shape sum isn't as horrible as it sounds and looks, though! The nice thing about the area of stuck-together rectangles is that you get the same answer if you split the rectangle into its individual pieces and then add the answers together. Even better, it doesn't matter *how* you split the shape up, as long as it's into other rectangles – the areas of the pieces always add up to the area of the whole thing.

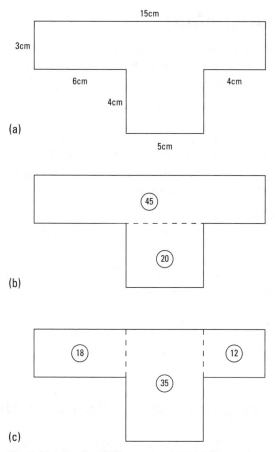

Figure 13-3: Dealing with compound rectangles.

In Figure 13-3, you can split the T-shape shown in (a) into (b) two or (c) three rectangles. The first way, shown in (b) gives you a long rectangle across the top with an area of 45cm² (15cm × 3cm), and a smaller rectangle below it with an area of 20cm² (5cm × 4cm), making a total of 65cm².

The second way, shown in (c), gives you a rectangle on the left with an area of 18cm² (3cm × 6cm), one on the right with an area of 12cm² and one in the middle that's 5cm wide and 7cm tall – its area is 35cm². In total, that's 65cm² – you get the same answer either way!

Laying plans

A very common exam question on the subject of size gives you a *map* or a *plan* of a shape and asks you to work out its area, its perimeter or both. Nine times out of ten, the shape will look just like the one in Figure 13-3 and you can approach it in exactly the same way:

> ✔ If you want the area, split it up into rectangles, find the area of each and add it up to get the answer.

> ✔ If you want the perimeter, work out the length of each side and then add all four sides together.

A slightly trickier version of this question may also be posed, though, which gives you a scale instead of labelling the sides for you. Fortunately, you can label the sides yourself: measure each of the sides and convert them to the 'real-life' measurements (read the 'Keeping to scale' section, earlier in this chapter, if you need a little help with the conversion).

 When you're given a scale, the smaller measurement refers to the map and the bigger one to real life. (Exceptions to this rule do exist, but as you won't see them in basic maths, don't worry about them.)

Attempting some plan questions

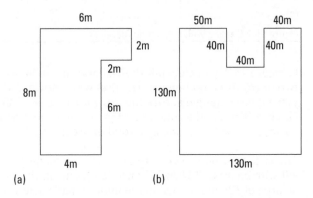

(a)   (b)

1. Diagram (a) in the above figure shows a plan of a room.

(a) What is the area of the room?

(b) What is its perimeter?

(c) Skirting board costs 55p per metre. How much would it cost to put skirting board around the room?

(d) Carpet tiles cost 45p per square metre. How much would it cost to carpet the room?

2. Diagram (b) in the above figure shows a map of a village green.

(a) What is the area of the green?

(b) What is the perimeter of the green?

(c) The local council decides to plant grass (which costs 25p per square metre) on half of the green, and flowerbeds (which cost £1.50 per square metre) on the other half. How much would this crazy plan cost?

Working through Review Questions

1. What is a sensible metric unit in which to measure:

(a) The length of a car?

(b) The distance between two cities?

(c) The width of a pencil?

2. What is a sensible imperial unit in which to measure:

(a) The size of a TV screen?

(b) The length of a marathon?

(c) The length of a barge-pole?

3. Convert:

(a) 163 centimetres to metres

(b) 4.6 kilometres to metres

(c) 300,000 millimetres to metres

(d) 500 miles to kilometres (1 mile = 1.6km)

(e) 9 inches to centimetres (1 inch = 2.5cm)

(f) 42 kilometres to miles (1 mile = 1.6 km)

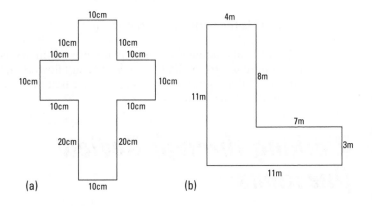

(a)　　　　　　　　　　(b)

4. Look at the shapes in the above figure:

(a) What is the perimeter of shape (a)?

(b) What is the area of shape (a)?

(c) What is the perimeter of shape (b)?

(d) What is the area of shape (b)?

5. What is the volume of a box if its dimensions are:

(a) 300cm long, 150cm high, 200cm wide?

(b) 5cm long, 10cm high, 20cm wide?

(c) 10cm long, 2cm high, 50cm wide?

6. I've got a drawer that's 30cm wide, 15cm high and 50cm deep. How many of the following things could I put in it?

(a) Kitchen timers that are 5cm wide, 3cm high and 5cm long.

(b) Calculators that are 7cm wide, 1cm high and 15cm long.

(c) Dice that are 1cm wide, 1cm high and 1cm long.

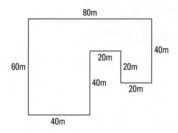

7. The above figure shows a plan for an ice-skating rink.

(a) What is the area of the rink?

(b) What is the perimeter of the rink?

(c) A Zamboni can brush 300 square metres of ice in a minute. How long would it take to brush the whole rink?

Checking Your Answers

Unit of length questions

1. (a) An athletics track is 400 *metres* all the way round.

(b) A door is about 200 *centimetres* high.

(c) London and Edinburgh are about 650 *kilometres* apart.

(d) A ruler is about 30 *centimetres* long.

Metric units of length always have the word 'metre' in them somewhere.

2. (a) This book is about 11 *inches* long.

(b) A door is about 7 *feet* high.

(c) London and Birmingham are about 100 *miles* apart.

(d) A football pitch is about 100 *yards* long.

Length, distance and perimeter questions

1. (a) 8cm   (b) 180cm (or 1.8 metres)   (c) 68cm

2. 2000 ÷ 25 = 80.

3. 20 + 5 + 15 + 15 + 10 + 40 = 105km.

4. 18 × 400 = 7,200m, or 7.2km.

5. 70 + 70 + 120 + 120 = 380m.

Conversion and scale questions

1. (a) 35cm   (b) 100,000m   (c) 4cm   (d) 15m

2. (a) 40cm   (b) 40 inches   (c) 5 miles   (d) 12,800m

3. (a) 4cm × 50,000 = 200,000cm = 2,000m = 2km.
   (b) 16 km = 16,000m = 1,600,000cm; 1,600,000cm ÷ 100,000 = 16cm.
   (c) 160km = 160,000m = 16,000,000cm; 16,000,000cm ÷ 1,600,000 = 10cm.

Volume and capacity questions

1. (a) 30 × 40 × 50 = 60,000cm^3.
   (b) 20 × 5 × 30 = 3,000cm^3.
   (c) 10 × 10 × 10 = 1,000cm^3.

2. The bigger cuboid has a volume of 4 × 5 × 10 = 200m^3. The smaller one has a volume of 2 × 3 × 5 = 30m^3. Altogether, that's 230m^3.

3. (a) $200 \times 50 \times 75 = 750{,}000 \text{cm}^3$.

(b) You can fit 10 copies along the way, 10 up the way and five across the way. So, $10 \times 10 \times 5 = 500$.

(c) You can fit 20 along the way, 5 up the way and seven back. (Don't cut Rubik's cubes in half! That just makes them even more difficult.) So, $20 \times 5 \times 7 = 700$.

Area questions

1. (a) $12 \times 5 = 60 \text{cm}^2$ (b) $6 \times 2.5 = 15 \text{cm}^2$ (c) $10 \times 12 = 120 \text{cm}^2$

2. $100 \times 70 = 7{,}000 \text{m}^2$.

3. $600 \times 450 = 270{,}000 \text{km}^2$.

Plan questions

1. (a) If you split plan (a) horizontally: the top part is $6 \times 2 = 12 \text{ m}^2$ and the bottom part is $6 \times 4 = 24 \text{m}^2$, making 36m^2 altogether. Splitting it vertically, the left-hand rectangle is $4 \times 8 = 32 \text{m}^2$ and the right-hand rectangle is $2 \times 2 = 4 \text{m}^2$ – making 36m^2 again.

(b) The perimeter is $6 + 2 + 2 + 5 + 4 + 8 = 27 \text{m}$.

(c) 27 metres $\times$ 55p = £14.85.

(d) $36 \text{m}^2 \times 45 \text{p} = £16.20$.

2. (a) The left-hand part of (b) is $50 \times 130 = 6{,}500 \text{m}^2$. The right-hand part is $40 \times 130 = 5{,}200 \text{m}^2$. The middle part is $40 \times 90 = 3{,}600 \text{m}^2$. That makes $15{,}300 \text{m}^2$ altogether. (You could also do $(130 \times 130) - (40 \times 40) = 16{,}900 - 1{,}600 = 15{,}300 \text{m}^2$.)

(b) $50 + 40 + 40 + 40 + 40 + 130 + 130 + 130 = 600 \text{m}$

(c) Half of the green is $7{,}650 \text{ m}^2$. So, $7{,}650 \times 25 \text{p} = £1{,}912.50$; $7{,}650 \times £1.50 = £11{,}475$. Altogether, the gardening costs £13,387.50. They could put in a petanque pitch for that kind of money!

Review questions

1. (a) metres (possibly centimetres)   (b) kilometres
    (c) millimetres (possibly centimetres).

2. (a) Inches (b) Miles
    (c) Feet. But I wouldn't touch imperial units with a ten- . . . oh.

3. (a) 1.63m   (b) 4,600m   (c) 30m
    (d) 800km   (e) 22.5cm   (f) 26.25 miles

4. (a) 140cm   (b) 500cm^2 (c) 44cm   (d) 65cm^2

5. (a) 9,000,000cm^3 (actually the same as 9m^3)
    (b) 1,000cm^3 (c) 1,000cm^3

6. (a) $6 \times 5 \times 10 = 300$ (b) $4 \times 15 \times 3 = 180$ (c) 22,500

7. (a) You need to work out that the narrow bit at the top is
$60 - 40 = 20$m tall. If you split it into vertical rectangles, you
get $2,400 + 400 + 800 = 3,600$m^2.

(b) 320m   (c) 12 minutes

Chapter 14

Sharpening Your Knowledge of Shapes

*I*n basic maths, you'll be pleased to hear, you don't need to know much about shapes. You just need to be able to:

✔ Recognise, name and draw a few basic shapes.

✔ Understand and measure angles.

✔ Recognise symmetry.

In this chapter, I take you through what you need to know about those topics, and also introduce you to some ways of drawing three-dimensional shapes.

Showing Off Your Shape Vocabulary

As a basic mathematician, you need to know about two kinds of shapes: *two-dimensional* (or 2D) shapes, which are flat (like a square and a circle) and *three-dimensional* (3D) shapes, which come out of the page (like a cube or a sphere). Think about the cinema: 2D films look like they're flat on the screen, but films in 3D jump out at you.

Recognising 2D shapes

Figure 14-1 shows the 2D shapes you need to know about:

- ✔ The *square* (a): four equal sides, right-angles at the corners.

- ✔ The *rectangle* (b): like a square, but the sides aren't necessarily equal.

- ✔ The *triangle* (c): three sides of any length. Triangles can be *equilateral* (which means all three sides are the same length), *isosceles* (if two of the sides are the same length) or *scalene* (the sides are all different lengths).

- ✔ The *circle* (d): a round figure consisting of points which are all the same distance from a fixed centre.

(a)　　　　(b)　　　　(c)　　　(d)

Figure 14-1: Identifying the 2D shapes you need to know: (a) the square; (b) the rectangle; (c) the triangle; (d) the circle.

While you need to be able to recognise and name all of these shapes, the most interesting ones for basic maths are the square and the rectangle because you can do easy sums to work out their area and perimeter (see Chapter 13 for the low-down on calculating perimeters).

You may also want to know about shapes with more sides. They're named after the old Greek word for the number of sides, followed by '-gon', which is Greek for 'corners'. For simple shapes, the number of corners is the same as the number of sides. Here are some examples of multi-sided shapes:

- ✔ Penta- is Greek for 5, and a 5-sided shape is a *pentagon*.

- ✔ Hexa- is Greek for 6, and a 6-sided shape is a *hexagon*.

- ✔ Octa- is Greek for 8, and an 8-sided shape is an *octagon*.

- ✔ Deca- is Greek for 10, and a 10-sided shape is a *decagon*.

Spotting 3D shapes

Figure 14-2 shows five different 3D shapes:

- ✔ The *cube* (a): has right angles at all of the corners and all of its sides are the same length (like dice).

- ✔ The *cuboid* (b): has right angles at the corners but the sides are different lengths (like shoe boxes).

- ✔ The *pyramid* (c): looks like a pyramid, hence the name; has a square on the bottom and four triangles on the sides.

- ✔ The *cone* (d): looks like the wafery bit of an ice-cream cone (hence the name).

- ✔ The *sphere* (e): ball-shaped.

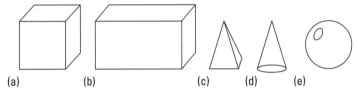

(a)   (b)   (c)   (d)   (e)

Figure 14-2: Identifying the 3D shapes you need to know: (a) the cube; (b) the cuboid; (c) the pyramid; (d) the cone; (e) the sphere.

You just need to be able to recognise these 3D shapes. For basic maths you'll only be asked to work out sums with two of them: the cube and the cuboid. Chapter 13 covers everything you need to know about finding the volume of these shapes.

Attempting some shape vocabulary questions

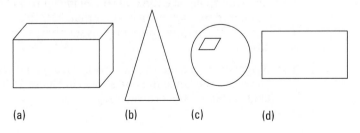

(a)   (b)   (c)   (d)

1. Match shapes A, B, C and D shown in the above figure with their name.

(a) Rectangle   (b) cuboid   (c) sphere   (d) triangle

2. Draw a picture of (it doesn't have to be perfect!):

(a) A square   (b) a cube   (c) a circle   (d) a pyramid

3. What is the difference between a square and a rectangle?

4. How many faces do the following shapes have?

(a) A cube   (b) A cuboid   (c) A pyramid

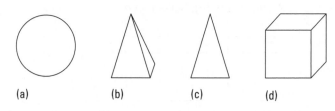

(a)   (b)   (c)   (d)

5. Look at the above figure. What is each shape called?

Turning the Corner with Angles

An *angle* is a fancy name for a corner. In this section, I show you how to measure angles – that is, to say precisely how sharp a corner is – and to answer questions about the size of angles in various shapes.

Before I do that, you need to know a few bits of vocabulary:

- ✔ An *acute* angle, like the one shown in Figure 14-3(a), is smaller than a right angle – less than 90°. I like to think of 'a cute little puppy' to remind myself that 'acute' means 'little'.

- ✔ A *right angle* is a nice square corner like you'd see on a door or a shelf, like the one shown in Figure 14-3(b). This angle is 90° on the protractor.

✔ An *obtuse* angle is one that's bigger than a right angle but less than two right angles, like the one shown in Figure 14-3(c). It's between 90° and 180°.

✔ A *reflex* angle is one that's bigger than two right angles, as shown in Figure 14-3(d); it's between 180° and 360°. I remember this angle by thinking about a doctor hitting the outside of my knee to test my reflexes – the angle on the outside of my knee is a reflex angle.

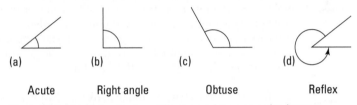

(a) (b) (c) (d)

Acute Right angle Obtuse Reflex

Figure 14-3: Identifying the four different angles you need to know.

Measuring angles

You measure an angle using a *protractor*. Figure 14-4 shows one of these see-through plastic semi-circular tools in action.

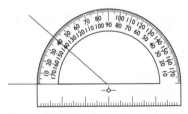

Figure 14-4: Using a protractor; this angle measures 40°.

To use a protractor to measure an angle that lies between two lines:

   1. Put the crosshairs in the middle of the protractor over the corner you want to measure.

2. **Turn the protractor so that the line marked zero lies over one of the lines defining the angle and the other line lies under the protractor.** It may help to turn the paper!

3. **Read off the number on the edge of the protractor where the second line emerges.** This is the angle!

Sometimes the line defining the angle isn't long enough to stretch out past the edge of the protractor. In that case, either make the line longer by drawing it with a ruler or make a sensible guess about where it would come out. Also, some protractors have two scales (one going clockwise and one going anti-clockwise); you must make sure you read off the correct one (the one that reads 0 on the first line).

Making sure your answers make sense is always a good idea. If your angle is acute but you measure it to be 140°, something has gone wrong!

But what if your angle is bigger than the protractor? Well, if it's bigger than 180°, you just need to measure the other side of the angle and then take your answer away from 360° (a full circle). The outside of the angle in Figure 14-4 is 360 – 40 = 320°.

Investigating angles in shapes

Working out the angles in a *regular* shape – one where all the sides are the same length and all of the angles are the same – is a little tricky. Fortunately, there's a recipe for working such angles out. Follow these steps:

1. **Count the number of sides and take away two.**

2. **Multiply the resulting number by 180°.**

3. **Divide the answer by the number of sides.** That's the angle at each corner.

So, for a hexagon – with six sides – you work out 4 × 180 = 720, and then divide 720 by 6 to get 120°. Each of the angles in a regular hexagon is 120°.

Attempting some angles questions

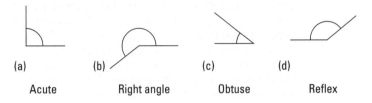

(a)   (b)   (c)   (d)

Acute   Right angle   Obtuse   Reflex

1. In the above figure, match angles (a), (b), (c) and (d) with the correct names.

2. Measure each of the angles marked in the above figure.

3. What size of angle would you find in:

(a) An equilateral triangle?   (b) A square?
(c) A regular hexagon?   (d) A rectangle?

Simplifying Symmetry

I mention the word 'regular' a few times in earlier sections of this chapter without really explaining why regular shapes are relevant. You spend more time in maths looking at squares and cubes than at trapezoids and parallelepipeds (don't ask) for a reason – and it's because of something called *symmetry*.

Symmetry is another one of those borrowed-from-Greek words: it translates, literally, as 'same-measure', but means something a bit different. A shape with symmetry is one that you can move around so it's somehow facing a different direction, but still looks the same.

You need to know about two types of symmetry: *reflective* symmetry, which means 'looking the same if you flip it over', and *rotational* symmetry, which means 'looking the same if you twist it around'. In this section, I show you both of these versions of symmetry in more detail, with pictures and everything!

Flipping things over

A shape has *reflective symmetry* if you can pick it up, flip it over and put it down so it still looks the same. Figure 14-5 demonstrates that the letters A, B, C, D and E all have reflective symmetry. So, you can flip A around sideways and you can flip the other four from top to bottom. H also has reflective symmetry – in two directions! You can flip it vertically or horizontally and it still looks like an H.

In Figure 14-5, I've drawn dashed lines through the middle of the letters with reflective symmetry. These lines are where you could put a mirror and the letter would look the same; get something shiny and try it. These are called *lines of symmetry* and you're often asked to count them or, in some exams, to draw them.

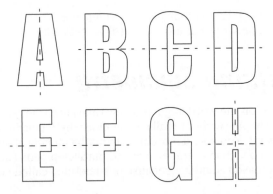

Figure 14-5: Looking at the lines of symmetry in some letters.

Turning things around

Rotational symmetry does what it says on the tin – if you can turn a shape around and it looks just the same, it has rotational symmetry. The *order* of rotational symmetry describes the number of different ways in which you can point the shape and have it look the same.

For instance, if you have a square, you can put it down with any of the four sides facing up, and all four arrangements would be the same – so a square has rotational symmetry of order four. A rectangle can only be turned in two ways (the right way up and upside-down) – so it has rotational symmetry of order two.

Not every shape has rotational symmetry; the letter W, for example, can only be put one way up without it looking like something else (an M or a 3 or an E, perhaps) – so the letter W has no rotational symmetry.

Figure 14-6 shows a few shapes and their orders of rotational symmetry.

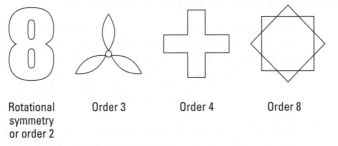

| Rotational symmetry or order 2 | Order 3 | Order 4 | Order 8 |

Figure 14-6: Looking at the rotational symmetry of four different shapes.

Attempting some symmetry questions

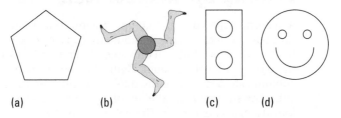

(a)   (b)   (c)   (d)

1. In the above figure, how many lines of symmetry are there in shapes (a), (b), (c) and (d)?

2. In the above figure, what is the order of rotational symmetry in shapes (a), (b), (c) and (d)?

3. How many lines of symmetry are there in:

(a) A square?   (b) A rectangle?   (c) A regular hexagon? (d) An equilateral triangle?

4. What is the order of rotational symmetry of the following symbols, and how many lines of symmetry do they have?

(a) 8   (b) Y   (c) %   (d) *   (e) K   (f) Z

Breaking Shapes Down

A basic maths exam question may ask you to draw 3D shapes. I'm not talking about using red and blue crayons and putting on special glasses so that your pyramids appear to rival the tombs of the Pharaohs. It doesn't even involve using perspective – which is a pity, because drawing nicely in 3D is one of the best skills a mathematician can have.

Instead, drawing in 3D concerns two particular ways of representing shapes. One method involves unfolding the surface of a shape and drawing the net that results; the other means you draw the way the shape looks from the front, the top and the side – the three elevations. In this section I show you how to think about nets and elevations.

Folding up shapes: Nets

A *net* is a shape you can fold up, origami-style, to make a three-dimensional shape without any gaps. Nets are really hard to visualise, but you only need to know a few shapes for the basic maths curriculum. Figure 14-7 shows nets of a few common shapes. Notice the difference between the net of a cube (six squares) and the net of a pyramid (a square with four triangles next to it). The net of a cone looks a bit like Pacman.

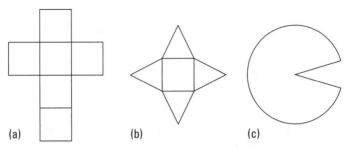

(a)　　　　　　　　(b)　　　　　　　　(c)

Figure 14-7: Looking at three different nets: (a) a cube; (b) a pyramid; (c) a cone.

Looking from all sides: Elevations

Architects like to show you what the thing they're building looks like from the front, back and sides, and sometimes even the top in case you need to pick it out on Google Earth or locate it from an aircraft.

Being able to sketch what an object looks like from the front and sides – and, conversely, to 'see' from those views what the object looks like in 3D – is unlikely to help you in an exam, but you may need to do it in an investigation if you take maths classes.

Understanding plans and elevations is also a useful step towards understanding 3D shapes for further study. Being able to visualise what's going on and having a good sense of up, back and sideways can make the shape part of GCSE maths more accessible and engaging.

To draw a plan – a top view – you imagine what you'd see if you looked down on your shape. What shape is the top? Would you see any lines? How big is the top? Then you sketch what you see. For the front elevation, you do the same thing, looking from the front. And for a side elevation, you draw what the side looks like.

When doing elevations and plans, don't take perspective into account. If you look at the front of a house where the roof slopes directly back, your eyes see the roof sloping inwards but the elevation shows the roof going straight up. Figure 14-8 provides a couple of examples of elevations and one plan.

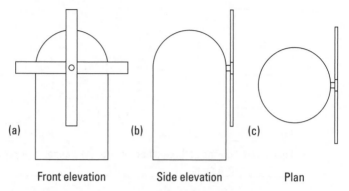

Front elevation   Side elevation   Plan

Figure 14-8: Drawing elevations: (a) the front; (b) the side; (c) the plan.

Attempting some nets and elevations questions

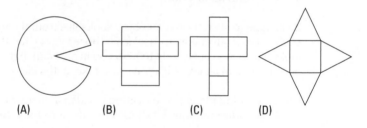

(A)   (B)   (C)   (D)

1. Look at nets A, B, C and D in the above figure. Which would fold up to make:

(a) A cuboid?   (b) a cube?   (c) a cone?   (d) a pyramid?

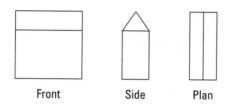

| Front | Side | Plan |

2. The above figure shows three elevations of an object. What is the object?

Working through Review Questions

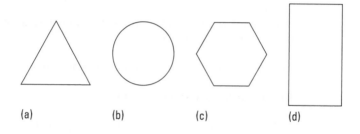

(a)   (b)   (c)   (d)

1. Name shapes (a), (b), (c) and (d) in the above figure.

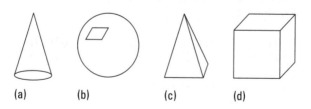

(a)   (b)   (c)   (d)

2. Name shapes (a), (b), (c) and (d) in the above figure.

3. Draw a rough sketch of:

(a) A pentagon   (b) a octagon   (c) a square   (d) a triangle.

4. What is the angle in each corner of the following regular shapes?

(a) An octagon (with 8 sides)?   (b) A decagon (with 10 sides)?
(c) An 18-sided shape?   (d) A pentagon (with 5 sides)?

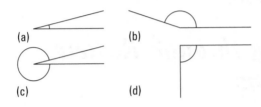

5. Label each of the angles in the figure above as an obtuse, reflex, acute or right angle.

6. Measure angles A, B, C and D in the above figure.

7. How many lines of symmetry do the following numbers have?

(a) 11   (b) 88   (c) 69   (d) 101
(e) 1008   (f) 906   (g) 48   (h) 318

8. What order of rotational symmetry do the following numbers have?

(a) 11   (b) 88   (c) 69   (d) 101
(e) 1008   (f) 906   (g) 48   (h) 318

9. Find a capital letter with:

(a) No rotational or reflective symmetry.
(b) Rotational symmetry but no reflective symmetry.
(c) Reflective symmetry but no rotational symmetry.
(d) Both reflective and rotational symmetry.

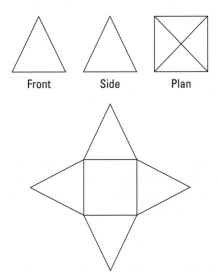

10. The above figure shows the net and elevation of a 3D shape. What is the shape?

Checking Your Answers

Don't worry too much if your sketches don't look quite like the sketches I provide here. Here are some important tips:

- ✔ Draw big. Don't be afraid to fill a quarter- or half-page with your diagrams.

- ✔ Get the main things down. If you're sketching, you don't need to measure all of your sides and angles, but try to make them look similar to what they should look like, and label them.

- ✔ If a diagram goes wrong, don't be afraid to scrunch up the paper and start again!

Shape vocabulary questions

1. (a) D – the rectangle   (b) A – the cuboid
   (c) C – the sphere   (d) B – the triangle.

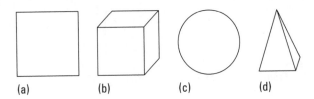

(a)   (b)   (c)   (d)

2. See the above figure.

3. A square has four equal sides that are all the same length, while a rectangle has a pair of sides that are one length and a pair of sides that can be a different length. (Both of them have right angles at each corner.)

4. (a) 6   (b) 6   (c) 5

5. (a) Circle   (b) pyramid   (c) triangle   (d) cube

Angles questions

1. (a) A right angle   (b) a reflex angle
   (c) an acute angle   (d) an obtuse angle

2. (a) 90°   (b) 225°   (c) 45°
   (d) 135° (answers within two degrees either side are fine)

3. (a) 60°   (b) 90°   (c) 120°   (d) 90°

Symmetry questions

1. (a) Five   (b) none   (c) two   (d) one.

2. (a) Five   (b) three   (c) two
   (d) no rotational symmetry.

3. (a) 4   (b) 2   (c) 6   (d) 3

4. (a) Rotational symmetry of order 2 and two lines of symmetry
   (b) No rotational symmetry, one line of symmetry
   (c) Rotational symmetry of order 2, two lines of symmetry
   (d) Five of each
   (e) Rotational symmetry of order two and no lines of symmetry.

Nets and elevations questions

1. (a) B   (b) C   (c) A   (d) D

2. A house – although I'm prepared to listen to more inventive answers!

Review questions

1. (a) Triangle   (b) circle   (c) hexagon   (d) rectangle

2. (a) Cone   (b) sphere   (c) pyramid   (d) cube

(a)   (b)   (c)   (d)

3. See the above figure.

4. (a) 135°   (b) 144°   (c) 160°   (d) 108°

5. A – acute, B – obtuse, C – reflex, D – right angle.

6. A – 15°, B – 150°, C – 345°, D – 90°.

7. (a) 2   (b) 2   (c) 0   (d) 2
   (e) 1   (f) 0   (g) 0   (h) 1

8. (a) 2   (b) 2   (c) 2   (d) 2
   (e) none   (f) 2   (g) none   (h) none

9. (a) F   (b) N   (c) B
   (d) O – although many others are acceptable answers!

10. A pyramid.

Part IV
Speaking Statistically

"I was a mathematician in a previous life – I specialised in multiplication."

In this part . . .

*I*f you've ever seen a PowerPoint presentation, you've seen a graph. In this part, I show you how to get accurate data off a graph, whether it's a line graph, a bar chart, a pie chart or something else.

It's also about basic statistics – how you find an average and a range of a set of numbers, and how you work out the probability of something happening.

Chapter 15

Mining Data (No Hard Hat Required)

● ●

In This Chapter

▶ Dealing with the different types of graphs and tables

▶ Working out what's wrong with some graphs

▶ Getting ready for more advanced graph questions

● ●

*R*eading tables and graphs is one of the most useful things I can teach you for the kind of everyday maths you may need in the workplace. Any time someone wants to communicate a load of numbers in a simple, easy-to-digest format, the odds are they'll use a graph or (sometimes) a number table.

A *graph* is simply a kind of picture that shows you numerical values somehow using the size bits of the picture. You need to know about several types of graph:

✔ Bar charts and pictograms, which usually look like a city skyline.

✔ Line graphs, which look like a zig-zag.

✔ Pie charts, which look like a pie, funnily enough.

You may also see *number tables*, which are just grids with numbers in, and possibly *tally charts*, which resemble a collection of fences.

In this chapter, I show you what each kind of graph and table means, how to get values from them (working out numbers from graphs and tables is sometimes called *data mining*) and what kind of mistakes people make with graphs. I also take you through some of the more involved questions that might come up in an exam; you often get asked a question about a graph that makes you think far beyond just reading off one value.

Reading Graphs and Tables

First things first! You probably need to be pretty good at two basic graph- and number-table reading skills before you can hope to correctly answer questions on this topic. The first is recognising which graph is which and the second – and possibly more important – is being able to read the right value from any kind of graph or table you're given. I help you develop both of these skills in this section.

Nailing down number tables

A number table does exactly what it says on the tin! It's a table, or grid, made up mainly of numbers. The numbers are arranged in rows and columns; each number is in a different *cell*, which in this context just means 'box'.

Name	Country	Age	Personal Best
C. Lebesgue	FRA	22	15.67
D. Hilbert	GER	25	16.75
R. Russell	GBR	19	14.95
J. Hamilton	IRL	24	17.14
J. Ferrari	ITA	27	17.02

Figure 15-1: A number table showing statistics for several athletes.

If you're familiar with spreadsheet programs, you may recognise the word *cell*, meaning a box with a number or a label in. The same concept applies in maths!

A number table usually has labels in the top row and the left-most column describing exactly what each of the cells represents (for instance, in Figure 15-1, the final column represents each athlete's personal best, and and the final row tells you all about J. Ferrari). Here's a recipe for how to find a particular value in a number table; let's imagine we have to find out R. Russell's age:

1. **Find the labels describing the thing you're looking for in the top row and the first column.** Here, you can see 'Age' in the top row and 'R. Russell' in the left-most column.

2. **Find the cell that's in the same column and row as the labels you found in Step 1.**

3. **The value in this cell is the number you're looking for.** In this case, it's 19.

Attempting some number table questions

	Area (km²)	Length (km)	Volume (km³)
Windermere	14.8	16.8	314.5
Ullswater	8.9	11.8	223.0
Derwent Water	5.4	4.6	29.0
Bassenthwaite Lake	5.3	6.2	27.9
Coniston Water	4.9	8.7	113.3

1. Use the data in the above number table to answer the following questions:

(a) Which of the lakes has the largest area?

(b) What is the volume of Coniston Water?

(c) What is the length of Derwent Water?

	Height (m)	Relative Height (m)
Scafell Pike	978	912
Helvellyn	950	712
Skiddaw	931	709
Great Gable	899	425
Cross Fell	893	651
Pillar	892	348

2. Use the data in the above number table to answer the following questions:

(a) Which of the mountains has the greatest height?

(b) What is the height of Great Gable?

(c) What is the *relative height* of Helvellyn?

Totting up tally charts

You can use a tally chart, like the one shown in Figure 15-2, if you're doing some kind of survey and want to record data quickly and efficiently. The idea is to organise people into groups of five as you go along to make them easier to count.

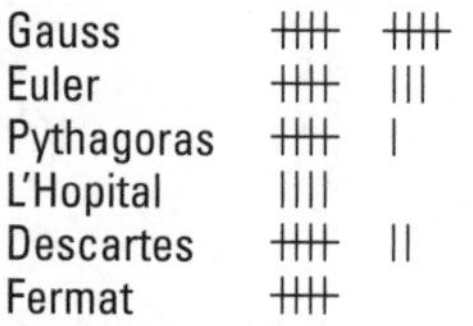

Gauss	十十 十十
Euler	十十 |||
Pythagoras	十十	
L'Hopital	||||
Descartes	十十 ||
Fermat	十十

Figure 15-2: A tally chart showing various mathematicians.

If an exam question concerns data in a tally chart, you'll probably be told that someone has done a survey (the example in Figure 15-2 is on their favourite mathematician of all time) and you have to find the number of people who chose a particular option, let's say Descartes. Follow these steps:

1. **Find the option you're interested in in the left-hand column.**

2. **Count the number of crossed-out fences to the right of your option.** Here, there's just one.

3. **Multiply this number by five.** In this example you get $1 \times 5 = 5$.

4. **Count the number of left-over marks, and add that on to the answer from Step 3.** Here, there are two extra marks, making a total of 7.

Attempting some tally chart questions

Sensible	⊪⊪ ⊪⊪ ⊪⊪ II
Silly	⊪⊪ ⊪⊪ III
Slightly Silly	⊪⊪ ⊪⊪ ⊪⊪ IIII

1. Use the data in the above tally chart to answer the following questions:

(a) Which is the most popular party in this survey?

(b) How many people would vote for the Slightly Silly party?

(c) Which party would 17 people vote for?

Under 18	⊪⊪ ⊪⊪ ⊪⊪ ⊪⊪ ⊪⊪
18-34	⊪⊪ ⊪⊪ ⊪⊪ II
35-64	⊪⊪ ⊪⊪ IIII
65 and over	⊪⊪ ⊪⊪ ⊪⊪ I

2. Use the data in the above tally chart to answer these questions:

(a) Which age group was least frequently seen at the cinema?

(b) How many people in the 35–64 age group went to the cinema?

(c) Which age group had 25 people in it?

Picking off pictograms

Figure 15-3 is an example of a *pictogram*. You may be able to figure out how it works from just looking at the diagram. You use a pictogram to compare the values of several different things, usually things you can count, and usually values that aren't very big. You don't want to be drawing hundreds of little pies!

The idea behind a pictogram is that you show the values you want to represent using pictures; in Figure 15-3, I use pies. Each little picture represents a certain number of whatever you're counting. The *key*, a little information box by the graph, tells you exactly how many. In this example, one pie represents four customers.

To find out the value of any given category in the graph, follow these steps:

1. **Find the category you're interested in.**

2. **Count how many of the little pictures there are next to the category name.** Ignore any part-pictures (like the last one in 'Blackberry', for example) for the moment.

3. **Find the number of things each picture represents, using the key.**

4. **Multiply your answers from Steps 2 and 3 together and write the answer down.**

5. **If you have a part-picture at the end, work out what fraction of a whole picture it is.** Normally, you're left with a quarter, a half or three-quarters.

6. **Find that fraction of your answer from Step 3.**

7. **Add the answer from Step 6 to the answer you wrote down in Step 4.** You're done!

You can see from Figure 15-3 that 16 people liked cherry pies and just seven liked blackberry pies.

Apple	○ ○ ○ ◖	
Blackberry	○ ◖	
Cherry	○ ○ ○ ○	**Key**
Plum	◖	○ 4 pies

Figure 15-3: A pictogram showing favourite pie flavours.

Attempting some pictogram questions

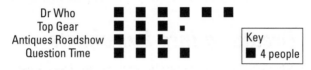

1. The pictogram above shows the results of a survey into people's favourite TV programmes. Use the data in the pictogram to answer the following questions:

(a) How many people's favourite TV show was *Dr Who*?

(b) Which show was the least popular?

(c) How many more people liked *Top Gear* than *Antiques Roadshow*?

Ford Focus	⁰⁰ ⁰⁰ ⁰⁰ ⁰⁰ ⁰	
Vauxhall Corsa	⁰⁰ ⁰⁰ ⁰⁰ ⁰	
Ford Fiesta	⁰⁰ ⁰⁰ ⁰⁰	**Key**
Vauxhall Astra	⁰⁰ ⁰⁰	⁰⁰ 4 cars
Volkswagen Golf	⁰⁰	

2. The pictogram above shows the results of a survey into which car people drove. Use the data to answer these questions:

(a) How many people in the survey drove Vauxhall Astras?

(b) How many people were surveyed altogether?

(c) Which was the least popular brand of car?

```
France      ::  ::  :
Germany     ::  :
   Italy    ::  ::  ::  ::      Key
   Spain    ::  ::  ::   ·      :: 6 people
```

3. The pictogram above shows the results of a survey into people's holiday destinations. Use the data to work out:

(a) How many people went on holiday to Italy.
(b) What was the most popular holiday destination.
(c) How many more people went to Spain than to France.

Breaking into bar charts

A *bar chart* (like the one shown in Figure 15-4) is like a pictogram but looks a bit more professional. You use a bar chart in the same circumstances in which you'd use a pictogram – when you want to compare the values of several different things.

In a bar chart, however, you don't usually need to use fractions to find out how big a value is. Instead of *counting* things to find the value, with a bar chart you *measure* the bars.

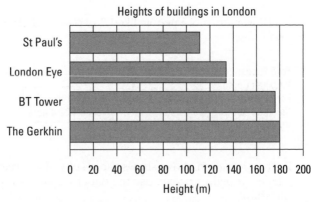

Figure 15-4: A bar chart showing the heights of various buildings in London.

Figure 15-4 shows the heights of several buildings in London. Imagine you want to know how tall the London Eye is. Now follow these steps:

1. **Find the label of the thing you're looking for.** Here, the London Eye is second from the bottom.

2. **Find the far end of the bar.** In this case, that means the right-hand end, but bar charts can also go up–down instead of left–right.

3. **Read the value on the numbered axis that's level with the end you found in Step 2.** That's your answer. Here, it's somewhere between 120 and 140 metres. It's closer to 140, so you might estimate 135 metres (and you'd be right).

You may also see *multiple* bar charts, as used in the 'Attempting some bar chart questions' section below. These show several different categories of related information at once. This multiple bar chart, for example, shows the exam results for three fictional schools over a few years. The results for any given year are grouped together, and each of the schools is shaded differently. (If this book was in colour, I'd represent the schools using red, blue and yellow, but you can't have everything.)

You read a multiple bar chart in much the same way as you do a single bar chart. The only difference is that you have to make sure you pick the right bar before you begin. You do so by looking at the key somewhere near the graph. It's usually a little box that tells you what each kind of shading (or colour) means.

Bar charts can sometimes be presented misleadingly. In a reasonable graph, the vertical axis should start at zero, but sometimes (either to save space or to cause mischief) people start it some way up. In the multiple bar chart used in the 'Attempting some bar chart questions' section, the vertical axis starts at 40 to save space. (See the 'Investigating Faulty Graphs' section, later in this chapter, for more on dealing with misleading or incorrect graphs.)

Attempting some bar chart questions

1. Use the data in Figure 15-4 to answer the following questions:

(a) Which is the tallest building listed?

(b) How much taller is the London Eye than St Paul's?

(c) How tall is the BT Tower?

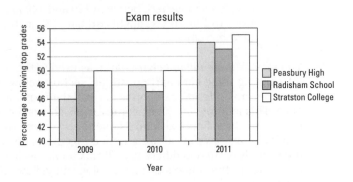

2. Use the data in the above multiple bar chart to answer these questions:

(a) What percentage of Peasbury High students achieved top grades in 2010?

(b) True or false: Stratston College achieved the best grades in all three years.

(c) True or false: only one of the schools improved every year.

Toeing the line (graph)

You normally use a line graph for one of two things:

   ✔ To show the changes in a value over time.

   ✔ To show how one value depends on another (for instance, if you know how many goals a footballer scored last season, how much is he worth in your Fantasy Football team?).

Figure 15-5 is an example of a line graph. You can see that, as well as the jagged line across the middle of the graph, it also has a straight line with ticks and numbers on (an *axis*) across the bottom and another up the side.

The plural of axis is *axes*. Confusing but true!

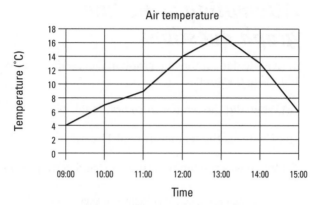

Figure 15-5: A line graph showing air temperature over a six-hour period.

Here's how you read a value from a line graph; let's consider how to work out the temperature at 10 a.m.:

1. **Work out which axis you're interested in.** The information you have is a time, so you look for the time axis, which is along the bottom.

2. **Find the value you're given on that axis.**

3. **Draw a straight line (either directly up or directly to the right) until you get to the line of the graph.** In this case, you go up.

4. **Draw a straight line directly to the other axis.** Here, you draw to the left.

5. **Read the value off this axis.** This is your answer! The temperature at 10 a.m. was 7°C.

You also need to be able to deal with *multiple line graphs*. These are, not surprisingly, just like normal line graphs but with extra lines. See the multiple line graph used in Question 2 in the 'Attempting some line graph questions' section below.

You follow exactly the same recipe for reading multiple line graphs as you do for normal line graphs. The only exception is that you need to start by reading the key to see which line you have to read. The lines may be in different patterns, widths or colours to show the different things being measured.

Attempting some line graph questions

1. Use the data in the line graph in Figure 15-5 to answer the following questions:

(a) When did the temperature first reach 14°C?

(b) What was the temperature at 11 a.m.?

(c) When was the temperature 13°C and getting cooler?

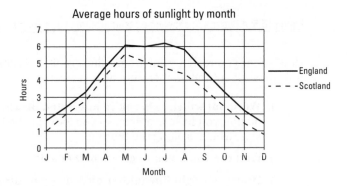

2. Use the data in the above multiple line graph to answer these questions:

(a) In which month did Scotland experience its highest average amount of sunlight?

(b) True or false: England received more hours of sunlight than Scotland in every month.

(c) For how many months of the year did England receive more than four hours of sunlight, on average?

Drawing perfect pie charts

You use a *pie chart* like the one shown in Figure 15-6 when you want to show the *proportions* of a total, instead of the actual value. The bigger the slice of pie, the more important that sector is. In Figure 15-6, which shows the proportions of people in a class with various eye-colours, you can see that the biggest proportion has blue eyes (although less than half) and the smallest has green eyes.

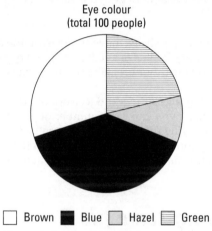

Eye colour
(total 100 people)

☐ Brown ■ Blue ☐ Hazel ▤ Green

Figure 15-6: A pie chart showing the proportion of people in a class with various eye-colours.

You *can* read values off a pie chart, though, if you know how many things or people there were in total. Say we want to know how many people have blue eyes. You just use the Table of Joy (Chapter 7 tells you everything you need to know about this wonderful tool) like this:

1. **Draw a noughts and crosses grid.** Label the columns as 'Degrees' and 'People' (or whatever you're counting), and the rows as 'Slice' and 'Circle'.

2. **Measure the angle of the slice you're interested in, at the middle of the circle.** The 'blue' slice has an angle of 108°. (If you don't know how to measure an angle, check out Chapter 14.)

3. **There are 360 degrees in a circle, so put 360 in the degrees/circle square. Then write the number of degrees in the slice (108) in the degrees/slice square, and the number of people there are altogether (100) in the people/circle square.**

4. **Shade the table like a chessboard and write down the Table of Joy sum. Then multiply the two numbers on the same colour squares and divide by the other one. So, 108 × 100 ÷ 360.**

5. **Work out the sum to get your answer.** I get 10,800 ÷ 360 = 30.

Incidentally, you don't always have to measure the angle. You can easily spot when a slice is a half (180°) or a quarter (90°) of the whole circle without breaking out the protractor.

You can use a *very* similar method to work out the total number of people if you know the number of people in one slice, just by changing the squares you fill in in Step 3. Instead of filling in the people/circle square, you fill in the people/slice square and do the sum as normal. Magically, you get the right answer again!

Attempting some pie chart questions

1. If you have a survey of 200 people, how big an angle would you need to represent:

(a) 50 people   (b) 150 people;   (c) 125 people.

2. A sales rep sells 90 items in a day and makes a pie chart to show which products she sold. How big a slice would she need to represent:

(a) 30 items;   (b) 20 items;   (c) 13 items.

Investigating Faulty Graphs

Graphs are a fantastic way of developing an intuitive understanding of how a set of numbers behaves, because your mind is much better at dealing with pictures than with numbers. However, you can encounter a big problem when trying to interpret graphs: if someone is feeling unscrupulous (or just doesn't realise what they're doing), they can quite easily draw a graph that's misleading.

As it happens, examiners are very fond of asking questions along the lines of 'what's wrong with this graph?'. Fortunately, in this section I take you through some of the things that can go wrong with a graph, and how to spot them.

Working out what's wrong

A graph can actually only go wrong in three ways:

- ✔ Problems with the axes.
- ✔ Problems with the way the graph is labelled.
- ✔ Problems with the way it's drawn.

In real life, whenever you encounter a graph it's worth having a quick check to see if someone's pulling the wool over your eyes; however, in an exam, the only time you really need to watch out is when a question specifically asks you what's wrong with a particular graph. Once you know the tricks (or mistakes – not everyone is deliberately trying to fool you, sometimes it's just incompetence), you can mentally run through them and see which one is the problem.

Problems with the axes

By far the most common problem with graphs – especially on election literature – is the 'broken axis' problem. Have a look at Figure 15-7 to see what I mean.

Figure 15-7(a), which the Silly Party would be quite likely to use in their leaflets, makes them look way ahead of the other two parties. If you look carefully, though, the vertical axis doesn't start at zero! If they'd drawn the graph honestly, it would look like Figure 15-7(b), where you can see that there's little difference between the parties.

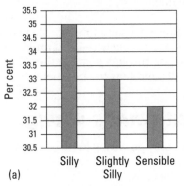

(a)

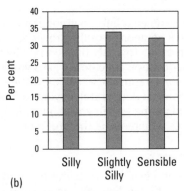

(b)

Figure 15-7: Recognising the broken axis problem: (a) the flawed graph; (b) the correct graph.

Similarly, some graphs make the mistake of spacing the axis values incorrectly. Figure 15-8 provides a couple of examples. In the bar chart in Figure 15-8(a), the values on the vertical axis are unevenly spaced – a gap of 100 is given between the first two values, then a gap of 200, a gap of 500 and so on. Figure 15-8(b) shows the correct bar chart.

In the line graph in Figure 15-8(c), the mistake is a bit trickier to spot: on the bottom axis, the months are unevenly spaced – it's labelled January, February, April and completely misses out March. Figure 15-8(d) is the correct version.

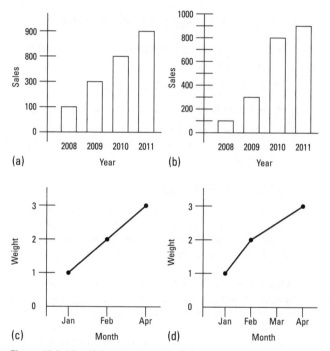

Figure 15-8: Identifying some more axis issues.

Lastly, consider another simple problem that also comes up once in a while: sometimes, the axes are just labelled incorrectly. If the vertical axis in Figure 15-8(a) was labelled 'Year' and the horizontal axis 'Sales' , clearly something would be wrong. Always check the labels!

Problems with the labels or key

Another of the many bits of a graph that can go wrong is the key, the little box somewhere on a graph that tells you what colour or pattern represents what.

You don't strictly *need* a key for every graph – simple bar charts and line graphs, for instance, make perfect sense without one – but almost all of the others do. The one possible exception is a pie chart; you can get away with labelling the sections and saying how many things there are in total in the title.

Look out for these key and label problems:

✔ No key or label is given (when one is needed to decipher the data).

✔ A colour or pattern is missing.

✔ In a pie chart, no indication of the total is given.

✔ The labels are mixed up (this can be hard to spot; you may have to check that the data matches up with the graph).

Problems with the picture

You may be extremely unlucky and be given a graph in a maths exam that's actually drawn incorrectly. One of the bars is the wrong size, one of the points on the line is in the wrong place, one of the slices is the wrong size . . . many possibilities exist.

Whilst it would be *very* unusual to see a badly-drawn graph in basic maths, if you can't see anything else wrong, checking that you'd draw the graph in the same way is worthwhile!

Chapter 16 gives you the lowdown on drawing graphs.

Attempting some faulty graph questions

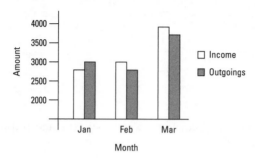

1. What is wrong with the bar chart in the above figure?

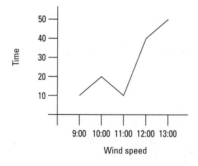

2. What is wrong with the line graph in the above figure?

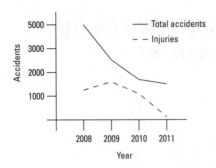

3. Which of the following is wrong with the line graph in the above figure?

A. A mistake on the vertical axis.

B. A mistake on the horizontal axis.

C. A mistake in the key.

D. Nothing; the graph is perfectly correct.

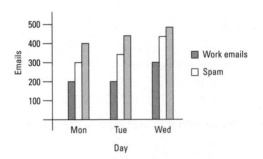

4. Which of the following is wrong with the bar chart in the above figure?

A. The axes are mis-labelled.

B. The key is wrong.

C. The bars are the wrong size.

D. Nothing; everything is as it should be.

Modes of transport

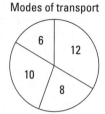

Cycling	‖‖ ‖‖ ‖	12
Walking	‖‖ ‖‖	8
Driving	‖‖ ‖‖	10
Other	‖‖ ‖	6

5. A researcher records information in the tally chart (a) shown in the above figure and uses the data to draw the pie chart (b). She has made one critical mistake in drawing the bar chart; what is it?

Grappling with Harder Graph Questions

As well as being able to read values off graphs, many maths tests ask you to do further sums with the answers you get. Typical questions include:

✔ **Finding the difference between two values**
Just find the values and subtract the lower one from the higher.

✔ **Finding the total of two or more values**
Simply find the values and add them up.

✔ **Finding the increase or decrease in a value between two times**
Read off the values and take them away.

✔ **Finding an average or range of values in a graph or table**
Write down the values and then work out the appropriate statistic (Chapter 17 covers everything you need to know about statistics). You may also need to find the difference between various averages.

✔ **Counting the number of values that fit certain criteria**
Find out which values fit the criteria – and then count them.

✔ **Finding or estimating what fraction or percentage of the total one observation makes**

Find the value in question and the total, and then work out your answer using the Table of Joy or your preferred percentages method (both described in Chapter 7) or a fraction method (covered in Chapter 5).

✔ **Spotting trends in a graph**
The statement 'Exam results improved every year', for example, can only be true if the graph never goes down.

Working through Review Questions

October

November

December

$\Box$ = 2 days

1. Richard kept track of his days off work in the last three months of the year and recorded his results in the pictogram above.

(a) How many days off did Richard take in November?

(b) What was his highest number of days off in any given month?

(c) How many more days did he take off in December than in October?

(d) How many days did he take off in total?

	Pass	C	B	A
Science	30	50	70	85
Maths	45	65	75	90
English	40	55	70	80

Table shows minimum mark needed to earn each grade.

2. An exam board used the number table above to decide which grades to award to candidates in different subjects.

(a) What was the lowest grade you'd have needed to get a B in Science?

(b) Josie got 75 marks in her Maths paper. What grade was she awarded?

(c) In which paper would you have needed at least 40 marks to pass?

Chicken	⊞ ⊞ ⊞ ⊞ |||
Pasta	⊞ ⊞ ⊞
Vegetarian	||||

3. A flight attendant keeps track of the meals ordered by the passengers on his flight using the tally chart shown above.

(a) How many people ordered the vegetarian meal?

(b) Which was more popular, chicken or pasta?

(c) How many meals did the passengers order altogether?

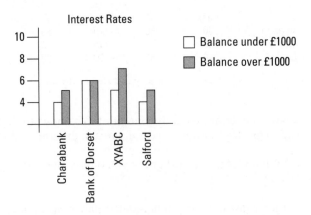

4. The bar chart above shows interest rates offered by several savings accounts.

(a) Which account offers the best interest rate if you're investing less than £1,000?

(b) What interest rate would you get if you invested more than £1,000 with the Bank of Dorset?

(c) What is wrong with this bar chart?

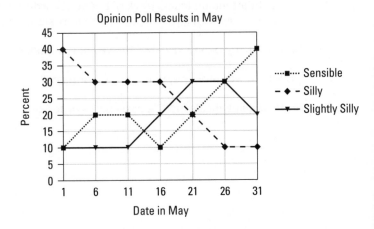

5. The multiple line graph above shows the opinion poll ratings of three political parties over a month-long election campaign.

(a) Which party was in the lead on 16 May?

(b) What was the highest percentage the Silly Party achieved over the month?

(c) How far ahead were the Sensible Party on 31 May?

(d) Which party gained the most support over the month?

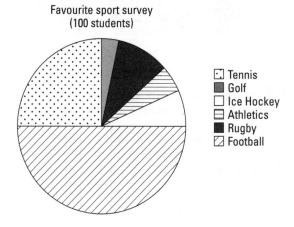

Favourite sport survey
(100 students)

Tennis
Golf
Ice Hockey
Athletics
Rugby
Football

6. The pie chart above shows the results of a survey about students' favourite sports.

(a) Which was the least popular sport in the survey?

(b) What percentage of students liked football best?

(c) How many students listed tennis as their favourite sport?

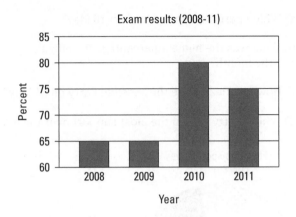

Exam results (2008-11)

7. The bar chart above shows a school's exam results over several years. Which of the following statements are true?

(a) The school's results improved every year.

(b) In the school's best year, 75% of students achieved top grades.

(c) Something is wrong with the vertical axis of the bar chart.

Checking Your Answers

With many graph questions, you're allowed some leeway in your answer. If, for example, a graph is printed fairly small, reading off values precisely can be difficult unless they fall exactly on a line. With that kind of question, rather than being expected to give 'one right answer', a range is acceptable.

Number table questions

1. (a) Windermere (14.8km^2)   (b) 113.3km^3 (c) 4.6km

2. (a) Scafell Pike (978m)   (b) 899m   (c) 712m

Tally chart questions

1. (a) Slightly Silly   (b) 19   (c) Sensible

2. (a) 35–64   (b) 17   (c) under 18

Pictogram questions

1. (a) 24   (b) *Antiques Roadshow* (obviously)
   (c) 2 (13 – 11)

2. (a) 8   (b) 56   (c) Volkswagen Golf

3. (a) 23   (b) Italy   (c) 4 (19 – 15)

Bar chart questions

1. (a) The Gherkin.
   (b) 24m (135 – 111); anything from 20 to 30 is fine.
   (c) 177m (anything between 170 and 180 is fine).

2. (a) 48.   (b) True.
   (c) True (Peasbury did. Radisham dropped from 2009/10, and Stratston had the same results in 2009 and 2010.)

Line graph questions

1. (a) 12:00   (b) 9°C   (c) 2 p.m.

2. (a) May   (b) True   (c) Six (April to September)

Pie chart questions

1. (a) 90°   (b) 270°   (c) 225°.

2. (a) 120°   (b) 80°   (c) 52°.

Faulty graph questions

1. The vertical axis doesn't start at zero.

2. The axes are labelled the wrong way round ('Time' should be on the horizontal axis).

3. A. There's a mistake on the vertical axis (it jumps from 2000 to 4000).

4. B. The key is wrong (it should say what the grey bar means!).

5. She hasn't labelled which section is which or given a key.

Review questions

1. (a) 5   (b) 10 (December)
   (c) 4 (10 − 6)   (d) 21 (6 + 5 + 10)

2. (a) 70   (b) B   (c) English

3. (a) 4   (b) Chicken (23 vs 15)   (c) 42 (23 + 15 + 4).

4. (a) Bank of Dorset (6%).   (b) 6%.
   (c) The vertical axis doesn't start at zero.

5. (a) Silly.   (b) 40%.   (c) 20% (40% − 20%).
   (d) Sensible. The Silly Party lost 30%, and the Slightly Silly party gained 10%.

6. (a) Golf   (b) 50%   (c) 25

7. (a) False. The results got worse from 2010 to 2011.
   (b) False. 2010 was the best year, when 80% of students got top grades.
   (c) True. The axis should start at zero.

Chapter 16

Grappling with Graphs

his chapter is all about drawing graphs, which is a useful skill to have even though it doesn't come up all that much in numeracy tests. This is because numeracy tests are almost always either computer based or multiple-choice.

However, an examiner may still ask you a few things about the process of drawing graphs, such as:

✔ Picking the correct kind of graph to use.

✔ Finding the angle for a sector of a pie chart.

✔ Finding fault with graphs.

In this chapter, I take you through all of these topics.

Picking the Right Graph

In basic maths, you need to know about four different kinds of graph:

✔ The *pictogram*, which uses pictures to show the size of each group.

✔ The *bar chart*, which has rectangles showing the size of each group.

✔ The *line graph*, which uses the height of a line to show the changes in a value over time.

✔ The *pie chart*, which uses the angle in the middle of a circle to show the relative sizes of several groups.

That's pretty much all you need to know! Pictograms and bar charts show the *actual* size of several groups; pie charts show the *proportional* size of the groups; and line graphs show the *evolution* of values over time.

Attempting some picking the right graph questions

1. Which graph would you use to depict:

(a) The change in house prices over a period of time?

(b) The proportion of people in a city belonging to different age groups?

(c) The cost of providing several different services?

(d) The different makes of cars parked on one street?

2. Sandy's boss has asked her to prepare a report featuring several graphs and tables. In each case she's suggested the wrong kind of diagram. Why are these diagrams not appropriate, and what would you suggest instead?

(a) A number table to show the comparative costs of four projects over 10 years.

(b) A pie chart to compare the expected time needed for the four projects.

(c) A pictogram to show the number of sales generated by each project.

Drawing Pie Charts

Working out the angle you need to have in the middle of a pie chart is the only time you need to work out a sum in this chapter! Luckily, you can use the Table of Joy to work out the angle you need (Chapter 7 gives you the lowdown on this marvellous grid). For instance, to work out the angle you need for a slice representing three out of ten people, you may follow these steps:

1. **Draw out a noughts and crosses grid, leaving plenty of room for labels.**

2. **Label the columns as 'Number' and 'Degrees' and the rows as 'Total' and 'Slice'.**

3. **Write down the numbers you know.** The number of things in the group you're interested (3) in goes in the number/slice square, the total number of things (10) goes in the number/total square and 360 goes in the degrees/total square because there are 360 degrees in a circle.

4. **Shade in the grid like a chessboard and write down the Table of Joy sum.** Multiply the two numbers on the same shaded squares and divide by the other number. The sum is $3 \times 360 \div 10$.

5. **Work out the sum.** The answer is the number of degrees in the slice. I get 108.

You can use the Table Joy to work out the total number of things, or the number of things in a slice, or just about anything pie-chart related, just by putting the right numbers in the right squares.

Attempting some pie chart questions

1. In a school year of 120 students, how big would the angle be in the centre of a pie chart representing:

(a) 60 of the students?   (b) 30 of the students?
(c) 10 of the students?   (d) 3 of the students?
(e) 13 of the students?   (f) 67 of the students?

2. What would the angle be at the centre of the slice of pie chart representing:

(a) 30 people out of 100?   (b) 45 people out of 135?

(c) 27 people out of 30?   (d) 27 people out of 36?

Finding Faults

Graphs can sometimes be misleading. Sometimes they're deliberately misleading (a politician may want to make statistics look better than they actually are) and sometimes they're incorrect by accident (people don't know any better). In this section I show you how to spot misleading and inaccurate graphs.

Identifying the broken axis error

The commonest problem with graphs, especially line graphs and bar charts, is for the vertical axis to be 'broken', that is, it doesn't start at zero. Not starting at zero can make differences between bars or lines look much bigger than they really are.

Figure 16-1 shows two graphs using the same data. Graph (a) has a vertical axis that doesn't start at zero, so it looks like the third bar is twice as big as the first. In reality, as shown in graph (b), the difference is much smaller!

The other major axis-related flaw can crop up if the numbers aren't evenly spaced. This is a hard one to spot!

Missing the key

Most graphs should have a *key* or a *legend* – a little box somewhere nearby that tells you what each element of the graph means. In a pictogram, you need to say how many things each picture represents; in almost every other graph, you need to say which colour or shading represents what.

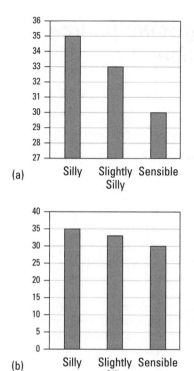

(a)

(b)

Figure 16-1: Reading a graph with (a) a broken axis and (b) a fixed axis.

In pie charts, you can sometimes get away with labelling the sectors rather than having a key; you also need to say somewhere how many things there are altogether.

If you can't see what some part of the graph represents, and no key is provided, then that's a *big* mistake!

Attempting some finding fault questions

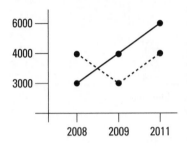

1. The graph above contains an error – or possibly more than one. Which of the following statements are true and which are false?

(a) There is an error in the legend.
(b) There is an error in the spacing on the *y*-axis.
(c) There is an error in the spacing on the *x*-axis.
(d) The *x*-axis should begin at 0.
(e) The *y*-axis should begin at 0.

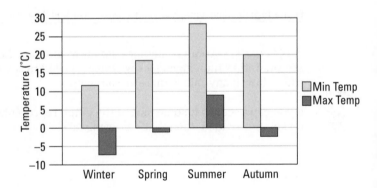

2. Above is another graph containing an error/ some errors. Which of the following statements are true and which are false?

(a) There is an error in the legend.
(b) There is an error in the spacing on the y-axis.
(c) There is an error in the spacing on the x-axis.
(d) The x-axis should begin at 0.
(e) The y-axis should begin at 0.

Working through Review Questions

1. Which graph would you use to represent:

(a) The number of points scored by two basketball players in a series of games?

(b) The relative proportions of pets owned by people in your street?

(c) Your bank balance over a period of several months?

(d) The favourite sports of people in a pub?

2. Which graph would you use to represent:

(a) The number of portable electronic devices owned by 50 people?

(b) The proportion of people who prefer different types of mobile phones?

(c) The number of people who prefer different types of mobile phones?

(d) The number of different electronic devices sold by a shop over several months.

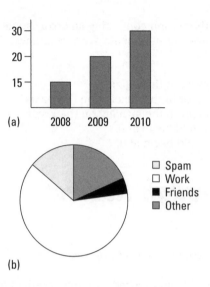

(a)

(b)

3. The bar chart (a) in the above figure contains at least one error. Identify which of the following statements are true and which are false:

(a) It's missing a legend.   (b) The bars are wrong.
(c) The horizontal axis is wrong.
(d) The vertical axis is wrong.

4. The pie chart (b) in the above figure contains at least one error. Identify which of the following statements are true and which are false:

(a) Some information is missing.   (b) The legend is wrong.
(c) The sizes of the slices are wrong.

5. A Twitter addict (who is entirely fictitious; any resemblance to myself is completely coincidental) sends 240 tweets in a day. He draws a pie chart to represent his prolific output. How big should the slices be to represent:

(a) Replies to real-life friends: 60 tweets?
(b) Retweets of news stories: 30 tweets?
(c) Complaints about deadlines or reports of writing success: 130 tweets?
(d) Silly jokes: 20 tweets?

Language	Speaker (millions)
Mandarin	840
Spanish	340
English	340
Hindi-Urdu	240
Arabic	220
Other	5220
Total	**7200**

6. The above table shows the number of people in the world who speak different languages. If you represent that data in a pie chart, how many degrees should the slices be for:

(a) Mandarin?   (b) Spanish?   (c) Hindi-Urdu?   (d) Other?

Checking Your Answers

In this chapter, more than most, some of the questions have more than one correct answer. If you get a different answer to me and can justify it to a friend, that's totally okay!

Picking the right graph questions

1. (a) A line graph.   (b) A pie chart.   (c) A bar chart.
   (d) A pictogram (possibly a pie chart or a bar chart, depending on exactly what you wanted to show and how many cars there were).

2. (a) A number table is quite hard to read when it gets that complex, so a line graph or possibly a bar chart would be better.

(b) A pie chart compares proportions, not amounts; I'd use a bar chart here.

(c) A pictogram is great for small amounts of data, but hopefully sales are better than that! I'd use a bar chart again.

Pie chart questions

1. (a) 180°   (b) 90°   (c) 30°   (d) 9°
   (e) 39°   (f) 201°

2. (a) 108°   (b) 120°   (c) 324°   (d) 270°

Finding fault questions

1. (a) True – the legend is missing.
(b) True – it has 3,000, 4,000, 6,000 evenly spaced.
(c) True – it jumps from 2009 to 2011.
(d) False – years don't need to begin at 0.
(e) True – values really should start at 0. Give yourself bonus points if you spotted that neither axis was labelled or that the 2008 label should be below the *y*-axis.

2. (a) True – the legend is the wrong way round.
   (b) False.   (c) False.   (d) False.   (e) False.

Review questions

1. (a) (Multiple) bar chart, or possibly a line graph.
   (b) Pie chart.   (c) Line graph.   (d) Pictogram or bar chart.

2. (a) A pictogram or bar chart.   (b) A pie chart.
   (c) A bar chart.   (d) A line graph.

3. (a) True – although it may not need one.
   (b) False.   (c) False.
   (d) True – it should have a label and be properly spaced.

4. (a) True – you don't know how many emails there are.
   (b) False – as far as you know.
   (c) False –as far as you know.

5. (a) 90°   (b) 45°   (c) 195°   (d) 30°

6. (a) 42°   (b) 17°   (c) 12°   (d) 261°

Chapter 17

Average Joe: Sussing Out Statistics

. . . .

In This Chapter

▶ Making sense of mean, median and mode

▶ Getting to grips with the grouped frequency table

▶ Working out the range in a set of numbers

. . . .

*T*his chapter is all about *statistics*. No, wait, come back! Let me at least explain what a statistic is: it's a number that tells you something about a set of values. That's all; nothing too geeky about that, is there? I mean, statistics *can* be geeky, but they don't have to be; saying 'a typical British male is about 175 cm tall' is just easier than reeling off a list of 30,000,000 heights. Actually . . . saying 'about 30,000,000 males live in the UK' is also a statistic – see, they're everywhere!

In this chapter I cover four main kinds of statistic – three different kinds of average (the mean, median and mode) and the range – all of which tell you different things about your set of data.

Asking about Averages

The word 'average' has a slightly different meaning in maths than in normal English. In both, an average is a number that describes a set of data or a list of numbers by telling you something about a typical or normal value for the list.

Actually (at least) three different kinds of average exist, all of which mean slightly different things:

✔ The *mean*, which is what you get if you divide the total of the numbers by how many things there are – the mean is a fair share of the total.

✔ The *median*, which is the 'number in the middle' – the median is a typical value.

✔ The *mode*, which is the most popular answer.

In this section, I show you how to work out each of these averages.

Meeting the mean

The *mean* means 'a fair share of the total'. I think it's the hardest of the three averages to work out – or, if you prefer, the meanest thing an exam question can ask.

Here's how you work out the mean of a list of numbers; take, for example:

6, 6, 7, 2, 2, 3, 4, 2

1. **Add up the numbers in the list to get the *total*.** So, $6 + 6 + 7 + 2 + 2 + 3 + 4 + 2 = 32$.

2. **Count how many numbers there are.** Here, there are eight.

3. **Divide the total by the count.** So, $32 \div 8 = 4$. That's your answer!

If you think about it, these steps are exactly what you take to split something – money, say – into fair shares: you put it all into one big pile and split it equally between the people involved!

Middling the median

The *median* means 'the thing in the middle'. Another word for the central reservation on a motorway is the 'median strip'.

To find the median of a list of numbers, such as

6, 6, 7, 2, 2, 3, 4, 2

follow these steps:

1. **Put the values in order.** This is a really important step that's easy to forget! I get 2, 2, 2, 3, 4, 6, 6, 7.

2. **Find the middle number or pair of values.** Here, 3 and 4 are the middle values (if the list had an odd number of values, only one value would be in the middle).

3. **If you see only one middle value, or the two values in the middle are the same, take that value as your answer.**

4. **If two different values are in the middle, find the mean of them.** Here $(3 + 4) \div 2 = 7 \div 2 = 3.5$.

Making the most of the mode

The mode is the easiest of the three averages to work out. It means 'the most common value in the list'. To work out the mode for a list of numbers such as

6, 6, 7, 2, 2, 3, 4, 2

just follow these steps:

1. **Count how many of each number there are.** Here, there are two 6s, one 7, three 2s, one 3 and one 4.

2. **Identify which number has the highest count.** That number is the mode; here, it's 2.

 If you can easily spot which number is the most common, you don't need to do the whole counting rigmarole, you can just write down the answer.

Attempting some averages questions

1. Kiran's scores on the first nine holes of the putting green were 2, 2, 5, 1, 7, 1, 4, 3 and 2.

   (a) What was the mode of his scores?

   (b) What was the median of his scores?

   (c) What was the mean of his scores?

2. George took a survey of the number of plant species in six different ecosystems. His results were 7, 4, 2, 9, 12 and 2.

(a) What was the mode of his results?

(b) What was the median number of plant species?

(c) What was the mean number of plant species?

Grouping Things Together

If mean, median and mode covered everything you need to know about averages, this would be a much shorter chapter! Unfortunately, you also have to understand another dimension – the *grouped frequency table* – which is about as appealing as it sounds. Figure 17-1 is an example of a grouped frequency table.

Number of letters	Frequency
1	1
2	3
3	9
4	5
5	7
6	4
7	2
8	0
9	2

Figure 17-1: A grouped frequency table.

You can see that this table is made up of two columns: the *value* column (usually something like 'number of. . .') and the *frequency* column, which means how many times that value shows up.

Tally charts, described in Chapter 15, are a very similar idea to grouped frequency tables.

The table in Figure 17-1 tells you about the number of letters in the words of a block of text. You can see that there's one word with one letter, three words with two letters, and so on up to two words with nine letters.

Making up the mean

Finding the mean of a grouped frequency table is one of the trickier questions you can be asked in basic maths, but if you can remember what the mean means and what the table is telling you, you can easily break down the steps. So, to find the mean length of the words in the sample of text analysed in Figure 17-1:

1. **Find the total number of things.** You do so by multiplying each value by its frequency and adding up the result. In this example, you're looking at letters, so: $(1 \times 1) + (2 \times 3) + (3 \times 9) + (4 \times 5) + (5 \times 7) + (6 \times 4) + (7 \times 2) + (8 \times 0) + (9 \times 2) = 1 + 6 + 27 + 20 + 35 + 24 + 14 + 0 + 18 = 145$.

2. **Work out how many piles you have to split things into by adding up the frequency column.** So, $1 + 3 + 9 + 5 + 7 + 4 + 2 + 0 + 2 = 33$.

3. **Divide the total by the count to get the mean.** Here, it's $145 \div 33$, which is (using a calculator) about 4.4.

If you have to work out a grouped frequency mean in basic maths, it'll be a much simpler sum than the one I describe here! This is the problem with using real data; the numbers can turn out horrible.

Finding the median

The median is a bit easier to find than the mean. In fact, all you have to do is figure out where the middle is. Here's how you find the median of the table in Figure 17-1:

1. **Find out the total frequency by adding up the frequency column.** Here, it's 33.

2. **If your answer is odd, add one, divide by two and then write down your answer.** Here, you have $34 \div 2 = 17$.

3. **If your answer is even, divide by two and write down the answer *and the next number*.** In a list of 50 numbers, you'd write down 25 and 26.

4. **Go through the frequency column, keeping a running total, until you get a number bigger than your answer in Steps 2 or 3.** Here, you'd say '1; 1 + 3 = 4; 4 + 9 = 13; 13 + 5 = 18, and I'm done.'

5. **Look at the value corresponding to the last number you added on.** This is your median! Here, it's 4.

What happens if you have two middle numbers that give you different values? Well, you'll be unlucky to see such a thing in basic maths, but all you need to do is take the mean of the two middle values.

Spotting the mode

The mode is by far the easiest statistic to find in a grouped frequency table. Showing you the recipe for it is almost a joke:

1. **Find the highest number in the frequency column.**

2. **Look at the value that corresponds to it.** That's the mode.

See? I told you it was easy.

Attempting some grouped frequency questions

Number of maths books owned	Frequency
0	4
1	8
2	14
3	9
4	4
5	1

1. The grouped frequency table above shows the results of a survey. Use the data to answer the following questions:

   (a) How many people responded to the survey?

   (b) What was the modal number of maths books owned by respondents?

(c) What was the median number of maths books owned by respondents?

(d) What was the total number of books owned by respondents?

(e) What was the mean number of books owned by respondents?

Hours spent online	Frequency
0	2
2	9
4	4
6	3
8	1
10	1

2. The grouped frequency table above shows statistics about Internet usage. Use the data to answer the following questions:

(a) How many people responded to the survey?

(b) What was the total number of hours spent online?

(c) What was the mean number of hours spent online?

(d) What was the modal number of hours spent online?

(e) What was the median number of hours spent online?

Ranging About

The *range* simply means the difference between the highest and lowest value in a set of numbers, and it gives you an idea of how spread out they are. For instance, if I look at a league table, the top team currently has 74 points; my team – eternally pinned to the bottom of the league – has 19. The range in the number of points is the difference between the numbers, 74 – 19 = 55 points.

Here's the recipe for finding the range:

1. **Find the highest value in the list or value column of the grouped frequency table.**

2. **Find the smallest value in the list or table.**

3. **Find the difference between the two (your answer from Step 1 take away your answer from Step 2).** This is the range.

Attempting some range questions

1. What is the range of prices at Giles's electronics shop if:

   (a) The cheapest item costs £4.50 and the most expensive costs £15.90?

   (b) The cheapest item costs £49 and the most expensive costs £999?

   (c) The cheapest item costs £15.49 and the most expensive costs £245?

2. What is the range of values in the following lists?

   (a) 11, 14, 2, 1, 0   (b) 8, 0, 0, 9, 1

   (c) 5, 1, 12, 13, 18, 33, 17   (d) 74.3, 44.5, 7.2, 35.9, 19.9

Working through Review Questions

1. A maths tutor keeps records of how many classes she gives each week for several weeks. Her results are: 12, 20, 16, 27, 12, 15 and 24.

   (a) For how many weeks did the tutor keep records?

   (b) How many classes did she give altogether?

   (c) What was the mean number of classes she gave per week?

(d) What was the modal number of classes she gave per week?

(e) What was the median number of classes she gave per week?

(f) What was the range of the number of classes she gave?

2. A rugby team scores the following number of points in each game of a tournament: 17, 13, 25, 21, 3, 17.

(a) How many games did they play?

(b) How many points did they score altogether?

(c) What was the mean number of points they scored per game?

(d) What was their median number of points?

(e) What was their modal number of points?

(f) What was the range of their number of points?

3. Jasmine records the number of minutes it takes her to commute each morning for two weeks. Her results are 35, 32, 23, 45, 25, 35, 30, 26, 19 and 35.

(a) How many mornings did Jasmine commute to work?

(b) How many minutes did she spend commuting over the two weeks?

(c) How much time is that converted into hours and minutes? (Chapter 9 covers everything you need to know about time.)

(d) How long was Jasmine's mean commuting time in minutes?

(e) What was Jasmine's median commute?

(f) What was the mode of Jasmine's commuting times?

(g) What was the range of Jasmine's commuting times?

4. Six judges are scoring a diving competition. To work out a competitor's score, you ignore the best and worst scores given by the judges, and find the mean of the remaining scores. The competitor with the highest score wins the competition.

The judges gave Darius's dive these scores: 3.6, 3.0, 5.7, 3.5, 1.7 and 5.7. They gave Jean's dive these scores: 2.8, 5.9, 2.7, 3.0, 3.6 and 5.9.

(a) What was Darius's score?

(b) What was Jean's score?

(c) Which of the two won the competition?

Number of employees	Frequency
1	20
2	9
3	2
4	2
5	1
6	1

5. Using the data in the above grouped frequency table answer the following questions:

(a) How many companies were surveyed?

(b) How many employees did they have in total?

(c) What was the mean number of employees per company? (Hint: you may want to cancel down a fraction here.)

(d) What was the median number of employees?

(e) What was the modal number of employees?

(f) What was the range in the number of employees?

Number of fillings	Frequency
0	8
1	25
2	7
3	5
4	4
5	1

6. The grouped frequency table above shows the number of fillings given by a dentist over the course of a week. Use the data to answer the following questions:

(a) How many patients were there altogether?

(b) How many fillings did the dentist give?

(c) What was the mean number of fillings per patient?

(d) What was the modal number of fillings?

(e) What was the median number of fillings?

(f) What was the range of the number of fillings?

Checking Your Answers

Many of these questions are more detailed than the ones you can expect in a maths test. That's deliberate: the questions walk you through the steps you need to take to answer the rest of the question.

Averages questions

1. (a) 2. The mode was 2, the most common result.

(b) 2. To find the median, you put the numbers in order and find the middle number. In order, Kiran's scores are: 1, 1, 1, 2, 2, 3, 4, 5, 7; the middle number is the fifth one, 2.

(c) 3. To find the mean, you add up the scores and divide by how many there are: adding up the scores gives 27, so the mean is 27 ÷ 9 = 3.

2. (a) 2. The mode is 2, the most common result.

(b) In order, the results are 2, 2, 4, 7, 9, 12. Because there are two 'middle' numbers (4 and 7), you have to take their mean, which is $(4 + 7) \div 2 = 11 \div 2 = 5.5$.

(c) 6. The total number of plant species is 36, and the mean is $36 \div 6 = 6$.

Grouped frequency questions

1. (a) 40. Add up the frequency column.

(b) 2. The most common number of books.

(c) 2. The 'middle numbers' would be the twentieth and twenty-first people in the list, who would each have 2 books.

(d) $(4 \times 0) + (8 \times 1) + (14 \times 2) + (9 \times 3) + (4 \times 4) + (1 \times 5) = 0 + 8 + 28 + 27 + 16 + 5 = 84$.

(e) $84 \div 40 = 2.1$.

2. (a) 20.

(b) $(0 \times 2) + (2 \times 9) + (4 \times 4) + (6 \times 3) + (8 \times 1) + (10 \times 1) = 0 + 18 + 16 + 18 + 8 + 10 = 70$.

(c) $70 \div 20 = 3.5$.

(d) 2.

(e) 2. The tenth and eleventh people both spent two hours online.

Range questions

1. (a) $15.90 - 4.50 =$ £11.40   (b) $999 - 49 =$ 950   (c) $245 - 15.49 =$ 229.51

2. (a) $14 - 0 = 14$ (b) $9 - 0 = 9$ (c) $33 - 1 = 32$ (d) $74.3 - 7.2 =$ 67.1

Review questions

1. (a) 7.

    (b) 126.

    (c) 126 ÷ 7 = 18.

    (d) 12. The most common answer.

    (e) 16. If you put the numbers in order, you get 12, 12, 15, 16, 20, 24, 27, and the middle number is 16.

    (f) 27 − 12 = 15.

2. (a) 6.

    (b) 96.

    (c) 96 ÷ 6 = 16.

    (d) In order, the points are 3, 13, 17, 17, 21, 25; both of the numbers in the middle are 17.

    (e) 17. Iit appears twice. (f) 25 − 3 = 22.

3. (a) 10 mornings.

    (b) 305 minutes.

    (c) 300 minutes is the same as five hours, so 305 minutes is 5 hours and 5 minutes.

    (d) 305 ÷ 10 = 30.5 minutes (or 30 minutes and 30 seconds).

    (e) In order, her times are 19, 23, 25, 26, 30, 32, 35, 35, 35 and 45 the two middle numbers are 30 and 32, so the median is (30 + 32) ÷ 2 = 31 – halfway between them.

    (f) 35 is the most common number in the list. (g) 45 − 19 = 26 minutes.

4. (a) To work out Darius's score, you ignore the worst score (1.7) and the best score (one of the 5.7s). You're left with 3.6, 3.0, 5.7 and 3.5, which add up to 15.8. Darius's score is $15.8 \div 4 = 3.95$.

   (b) For Jean's dive, ignore the 2.7 and one of the 5.9s to leave 2.8, 5.9, 3.0 and 3.6. These add up to 15.3. Jean's score is $15.3 \div 4 = 3.825$.

   (c) Darius had the higher score, so he won the medal.

5. (a) 35. The total of the frequency column.

   (b) $20 \times 1 + 9 \times 2 + 2 \times 3 + 2 \times 4 + 1 \times 5 + 1 \times 6 = 20 + 18 + 6 + 8 + 5 + 6 = 63$.

   (c) $63 \div 35 = 9 \div 5 = 1.8$.

   (d) 1. The eighteenth-largest company is the 'middle' one, and it has one employee.

   (e) 1. More companies have one employee than any other number.

   (f) $6 - 1 = 5$.

6. (a) 50. Add up the frequency column.

   (b) $0 \times 8 + 1 \times 25 + 2 \times 7 + 3 \times 5 + 4 \times 4 + 5 \times 1 = 0 + 25 + 14 + 15 + 16 + 5 = 75$.

   (c) $75 \div 50 = 1.5$.

   (d) 1. The most common number; 25 people needed one filling.

   (e) 1. The twenty-fifth- and twenty-sixth-highest number of fillings would be in the middle, and they'd both be in the '1' group.

   (f) $5 - 0 = 5$.

Chapter 18

What Are the Chances? Playing with Probability

. . . .

In This Chapter

▶ Working out the odds of something happening

▶ Doing experiments to work out probabilities

▶ Calculating probabilities for two things at once

. . . .

*P*robability, bizarrely, isn't part of the national numeracy curriculum, but when I rule the world, it will be. It's one of the most interesting and useful areas of maths (in my opinion) and a great way to practise the skills you learn elsewhere in this book.

So, let me be clear: although anything in this chapter is unlikely to come up in a numeracy test probability is a big part of GCSE and it's also useful for working out strategies for any game that involves chance (such as card and dice games, including poker and Monopoly).

Figuring Out the Odds

You hear, and probably use, probabilistic language all the time. You may describe something that could go either way as a coin-toss or a 50:50 chance. You may say 'there's a 30 per cent chance of rain tomorrow', or 'I've got a one in four chance of getting this promotion'. All of these things are *probabilities*.

More precisely, a *probability* is a number that describes how frequently you expect something to happen. Something that'll never, ever happen – February having 35 days, or throwing a normal die and getting a seven, or Scotland winning the football World Cup – these things have a probability of zero. Things that will definitely happen – somewhere in the UK experiencing rain this year, the winner of the men's 100 m in the Olympics being a man, me being annoyed by somebody on *Question Time* – these things all have a probability of one.

Living with likelihood

In-between zero and one is where the interesting stuff happens, though! Probability is always a number between 0 and 1: the closer the number is to 0, the less likely the thing (or *event*) is; the closer it is to 1, the more likely the event is. So, something with a probability of 0.99 is very likely to happen; something with a probability of 0.01 is very unlikely. And something with a probability of 0.5 is right in the middle – it's as likely to happen as it is not to happen . . . like a coin landing on heads.

So, how do you work out a probability? Doing so is quite easy. Let's say I want to know the probability of picking a red pen out of the jar on my desk. Here's what I do:

1. **Count the number of possible *positive outcomes*, meaning how many ways in which the event I'm interested in can happen.** In my jar are three red pens, so there are three positive outcomes.

2. **Count the number of *possible outcomes* , that is, how many possible events there are.** My pot contains nine pens altogether, including the red ones, so there are nine possible outcomes.

3. **Write the answer from Step 1 as the top of a fraction and the answer from Step 2 as the bottom. Cancel it down.** The resulting fraction is your answer. I get $\frac{3}{9} = \frac{1}{3}$, which is the probability of my pulling a red pen from my pot.

So, for example, if you want to know the probability of throwing a 6 on the geeky, 20-sided die I have on my desk, I'd say: 'I have one positive outcome (there's only one way to throw a 6) out of a possible 20 outcomes (there are 20 sides), so the probability is $\frac{1}{20}$.'

I've assumed here that life is *fair*. Obviously, life isn't really fair or I'd have a chauffeur and a mansion, but probability questions usually make up for it. Calling something *fair* means that all of the outcomes are equally likely; if you roll a die over and over and over again, you expect to see as many sixes as ones. The opposite of fair is *biased*; a biased coin may land on heads more often than tails or vice versa.

Being contrary

You can also work out the probability of something *not* happening. The process is very easy:

1. **Work out the probability of the event happening.**

2. **Take your answer away from 1.** That's the probability of the event not happening.

To work out the probability of *not* throwing a six on a normal, six-sided die, you find the event's probability (⅙) and then take that away from 1, making ⅚. That's the probability of you not throwing a six.

Attempting some probability questions

1. What is the probability (as a fraction) of:

(a) Tossing a head on a fair coin?

(b) Rolling a three on a fair die?

(c) Pulling an ace from a shuffled pack of cards?

(d) Rolling a seven on a fair die?

A die has six sides, unless you're told otherwise.

2. Forty-nine numbered balls are used for the National Lottery; you pick six numbers for your ticket before the lottery draw.

   (a) What is the probability that the first ball drawn is number 42?

   (b) What is the probability that the first ball drawn isn't number 42?

   (c) What is the probability that the first ball drawn is one of the six you chose?

   (d) What is the probability that the first ball drawn isn't one of the six you chose?

Running Experiments

A *statistical experiment* is a fancy name for something quite boring: trying something – a *trial* – over and over again to see (for instance) what the probability is. Other kinds of statistical experiments exist, but you don't need to worry about them until A-level maths.

Finding a probability

You find a probability from an experiment in a similar way to finding a fair probability (as described in the previous section). Let's say I'm keeping track of the days I go for a run, and in April I manage 12 runs in a 30-day period.

1. **Count the number of trials with positive outcomes.** Here, it's 12.

2. **Count the number of trials altogether.** Here, it's 30.

3. **Write the answer from Step 1 as the top of a fraction and the answer from Step 2 as the bottom.** This is your estimated probability. So, here that's $\frac{12}{30}$ or $\frac{2}{5}$.

You can also follow this process the other way! If you know a probability and how many trials you plan to run, you can figure out your *expected number of successes*. All you do is multiply the probability by the number of trials. So, if you're rolling a die 90 times, you'd expect to roll a three $\frac{1}{6} \times 90 = 15$ times.

If you throw a die 90 times in real life, you're not guaranteed to get exactly 15 threes; you may get a few more or a few less. But, on average, you'd expect to get 15.

Attempting some experiments questions

1.The probability of you winning a game in a casino is ³⁄₂₀. How many times would you expect to win if you played:

(a) 20 games? (b) 60 games?

(c) 100 games? (d) 240 games?

2. Below are the win records of some tennis players. Estimate the probability that each of them wins a game selected at random. Give your answer as a fraction.

(a) Andy won 60 games out of 80. (b) Boris won 45 games out of 50.

(c) Carlos won 12 games out of 40. (d) David won 12 games out of 60.

3. What results would you expect in the following experiments?

(a) You roll a fair die 120 times and count how many times you throw a five or a six.

(b) You toss a coin 100 times and count how many heads you get.

(c) You show a fake psychic (as seen on TV!) 100 cards, each selected at random. The psychic guesses the suit of each card. How many would you expect him to guess correctly?

(d) A train company's services run punctually 90 per cent of the time. If I use their services twice every weekday for four weeks, how many times would I expect to be delayed?

Try converting the percentage into a fraction.

Combining Events

Probability gets more complicated if you try to think about two things at once. For example, if you've ever tried to follow the last day of the football season, you'll know how confusing it gets when you're thinking, 'If Blackburn score again and West Brom concede, then Wigan will be safe. . .' and so on. Let me try to clear it up a bit!

In this section, I show you how to deal with two kinds of combination: the 'both or all' kind of event, where both things have to happen to count as a success; and the 'either or both' kind of event, where only one of the things has to happen.

Both or all events

A good example of a both or all event would be my friend Clyde getting a standing ovation at an open mic night. For that to happen he has to be in a good mood (the first event) and he has to play his signature song *Absolutely No-one* (the second event). If both of those things happen, the crowd is guaranteed to go wild. However, if he's feeling grumpy, the crowd will sense it and won't applaud much. And if he decides not to play *Absolutely No-one*, that's exactly who'll clap.

To work out the probability of a both or all event happening, here's what you do. I'm going to assume that Clyde is in a good mood with probability $\frac{9}{10}$ (he's a cheery bloke) and plays *Absolutely No-one* with probability $\frac{4}{5}$ (he likes to explore new musical directions once in a while). Follow these steps:

1. **Multiply the tops of your probabilities together.**
   So, $9 \times 4 = 36$. This is the top of your answer.

2. **Multiply the bottoms of your answers together.**
   That's $10 \times 5 = 50$. This is the bottom of your answer.

3. **Cancel the fraction down if you can.** So, $\frac{36}{50} = \frac{18}{25}$.
   This is the probability of Clyde receiving a standing ovation.

This example assumes that the events are *independent*, which means that the outcome of one event doesn't affect the other. If Clyde only plays *Absolutely No-one* when he's in a good mood that changes the sums a lot! Luckily, in all of the questions you're likely to come across, the events will be independent.

Either or both events

Either or both events are slightly different, and a little bit harder to work out. My favourite way to think of this sort of event is to take a pack of cards and find the probability of getting – for instance – a red card or a 10.

You can work out this probability in two ways. First is the long, boring way, which is to go through all 52 cards and decide whether each is a success (either red, or a 10, or a red 10) or not. If you did that, and I don't recommend it, you'd get 28 successes out of 52, so the probability is $^{28}/_{52}$ = $^{7}/_{13}$.

The second, and more mathsy, way of working out the probability is like this:

1. **Work out the probability of the first event.** In this case, half the cards are red, so it's $\frac{1}{2}$.

2. **Work out the probability of the other event.** There are four tens in the pack, so the probability of picking a 10 is $^{4}/_{52}$ or $^{1}/_{13}$.

3. **Find the probability of both things happening.** This is the situation explained in the preceding section. Here, that's $\frac{1}{2} \times \frac{1}{13} = \frac{1}{26}$.

4. **Add your answers from Steps 1 and 2 and then take away your answer from Step 3.** I get $\frac{1}{2} + \frac{1}{13} = \frac{15}{26}$, then $\frac{15}{26} - \frac{1}{26} = \frac{14}{26} = \frac{7}{13}$, which is the same as the answer you'd get if you went through all the cards individually!

The reason you take away the 'both' probability at the end is because otherwise you double-count some of the events – here, the red 10s.

Attempting some combining events questions

1. I roll a fair die and toss a fair coin. What is the probability that I:

   (a) Roll a six on the die?

   (b) Toss a head on the coin?

   (c) Roll a six and toss a head?

   (d) Roll a six or toss a head, or both?

2. I pull a card at random from a pack of cards. What is the probability that it's:

   (a) A jack?

   (b) A heart?

   (c) The jack of hearts?

   (d) A jack and/or a heart?

3. I have a bag containing the numbers 1 to 100, and I pull out a number at random.

   (a) What is the probability that my number is even?

   (b) What is the probability that my number is a multiple of 5?

   (c) What is the probability that my number is an even multiple of 5?

   (d) What is the probability that my number is a multiple of 10?

   (e) What is the probability that my number is a multiple of 2 and/or 5?

Working through Review Questions

1. What is the probability that I roll:

   (a) A 2 on a fair die?

   (b) An even number on a fair die?

   (c) A seven on a fair die?

   (d) A number below seven on a fair die?

2. A bag contains 90 balls, numbered 1 to 90.

   (a) What is the probability of pulling out legs 11?

   (b) What is the probability of pulling out an odd number?

   (c) What is the probability of drawing 10 or less?

   (d) On my bingo card are 15 numbers. What is the probability of the first ball out of the bag being on my card?

3. The probability of drawing a pair of cards in poker is about ⅙. How many times would you expect to get a pair if you played:

   (a) 6 hands?   (b) 60 hands?   (c) 120 hands?   (d) 180 hands?

4. Estimate the probability of the following football teams winning a game selected at random, given their records:

   (a) Notvery Athletic have won 3 games out of 24.

   (b) Dummies United have won 12 games out of 18.

   (c) Olympique Goldmedal have won 15 games out of 20.

   (d) Electri City have won 15 games out of 25.

5. If I toss a fair coin and roll a fair die:

    (a) What is the probability of getting a head and a 6?

    (b) What is the probability of getting a head *or* a 6, or both?

    (c) What is the probability of getting a head or a six, but *not* both?

    (d) What is the probability of getting neither a head nor a six?

6. If I pull a random card out of a regular, shuffled pack, what is the probability that I pull out:

    (a) The king of diamonds?

    (b) A black jack?

    (c) A picture card (king, queen or jack)?

    (d) A black picture card?

Checking Your Answers

This chapter is an opportunity to use your fractions skills. I cover fractions in detail in Chapter 5, so take a look if you need to. Fractions are much more versatile than decimals for probability sums, so getting in some practice here is a good idea!

Probability questions

1. (a) ½ – 'head' is one of two equally likely outcomes.

    (b) ⅙ – 'three' is one of six equally likely outcomes.

    (c) ⁴⁄₅₂ or ¹⁄₁₃ – there are four aces in a pack of 52 cards.

    (d) 0 – 'seven' isn't a possible outcome on a normal die.

2. (a) $\frac{1}{49}$ – '42' is one of 49 equally likely outcomes.

   (b) $\frac{48}{49}$ – there are 48 balls that aren't '42'.

   (c) $\frac{6}{49}$ – you have named six of the 49 possible outcomes.

   (d) $\frac{43}{49}$ – out of the 49 balls, you picked six; which means there are 43 balls left that you didn't pick.

Experiment questions

1. (a) $20 \times 3 \div 20 = 3$ (b) $60 \times 3 \div 20 = 9$

   (c) $100 \times 3 \div 20 = 15$ (d) $240 \times 3 \div 20 = 36$

2. (a) $\frac{60}{80} = \frac{3}{4}$ (b) $\frac{45}{50} = \frac{9}{10}$

   (c) $\frac{12}{40} = \frac{3}{10}$ (d) $\frac{12}{60} = \frac{1}{5}$

3. (a) The probability is $\frac{2}{6}$ (or $\frac{1}{3}$); the number of successful trials is $120 \times 2 \div 6 = 40$. (Alternatively, $120 \times 1 \div 3 = 40$ as well.)

   (b) The probability is $\frac{1}{2}$; the number of heads is $100 \times 1 \div 2 = 50$.

   (c) The probability is $\frac{1}{4}$; the number of correct guesses is $100 \times 1 \div 4 = 25$.

   (d) I make 10 journeys a week (two per weekday), so that's 40 journeys in four weeks. If $\frac{9}{10}$ of the services are on time, I'm on time $40 \times 9 \div 10 = 36$ times, and late four times.

Combining events questions

1. (a) $\frac{1}{6}$ (b) $\frac{1}{2}$ (c) $\frac{1}{6} \times \frac{1}{2} = \frac{1}{12}$ (d) $\frac{1}{6} + \frac{1}{2} - \frac{1}{12} = \frac{7}{12}$

2. (a) $\frac{1}{13}$ (b) $\frac{1}{4}$ (c) $\frac{1}{52}$ (d) $\frac{16}{52} = \frac{4}{13}$

3. (a) $^{50}\!/_{100} = \frac{1}{2}$ (b) $^{20}\!/_{100} = \frac{1}{5}$ (c) $\frac{1}{2} \times \frac{1}{5} = \frac{1}{10}$

(d) $\frac{1}{10}$ (actually the same numbers as for part (c)!)

(e) $\frac{1}{2} + \frac{1}{5} - \frac{1}{10} = \frac{6}{10} = \frac{3}{5}$

Review questions

1. (a) $\frac{1}{6}$ (b) $\frac{3}{6} = \frac{1}{2}$ (c) 0 (impossible!) (d) 1 (certain!)

2. (a) $\frac{1}{90}$ (b) $^{45}\!/_{90} = \frac{1}{2}$ (c) $^{10}\!/_{90} = \frac{1}{9}$ (d) $^{15}\!/_{90} = \frac{3}{18} = \frac{1}{6}$

3. (a) $6 \times \frac{1}{6} = 1$ (b) $60 \times \frac{1}{6} = 10$ (c) $120 \times \frac{1}{6} = 20$ (d) $180 \times \frac{1}{6} = 30$

In this example, you could also just divide by 6. However, the sum wouldn't work if the probability didn't have a 1 on top, so remembering to multiply by the probability is the best approach.

4. (a) $\frac{3}{24} = \frac{1}{8}$ (b) $^{12}\!/_{18} = \frac{2}{3}$ (c) $^{15}\!/_{20} = \frac{3}{4}$ (d) $^{15}\!/_{25} = \frac{3}{5}$

5. (a) $\frac{1}{2} \times \frac{1}{6} = \frac{1}{12}$ (b) $\frac{1}{2} + \frac{1}{6} - \frac{1}{12} = \frac{7}{12}$ (c) $\frac{7}{12} - \frac{1}{12} = \frac{1}{2}$

(d) $\frac{1}{2} \times \frac{5}{6} = \frac{5}{12}$ (alternatively, $1 - \frac{7}{12} = \frac{5}{12}$)

6. (a) $\frac{1}{52}$ (b) $\frac{2}{52} = \frac{1}{26}$ (c) $^{12}\!/_{52} = \frac{3}{13}$ (d) $\frac{6}{52} = \frac{3}{26}$

Part V
The Part of Tens

"It's not fair – he gets called out
more than any of us."

In this part . . .

*I*t wouldn't be a *For Dummies* book without a Part of Tens! In this part, I give you a few – well, $3 \times 10 = 30$ (or so) – pro tips on how to check your work, how to remember your number facts and some sneaky tricks for doing well!

Chapter 19

Ten (Or So) Ways to Check Your Work

In This Chapter

▶ Making sure you use all the information you're given

▶ Starting with a rough answer

▶ Drawing useful diagrams

I know what it's like when you're doing a big list of sums. You just want to get it done so you can get on with your life – dinner's on the table and your favourite programme starts in half an hour.

But please – take a few moments to check your work before you finish. Your dinner's probably too hot to eat right now and you can catch that programme on the Internet. Here are ten ways to give your work the quick once-over and avoid potentially embarrassing mistakes.

Making Sense

In almost every sum you do, making sure that what you've written down is a reasonable answer to the original question is worthwhile. Here are a few questions you can ask yourself about your answer:

  ✔ *Adding*: Is your answer bigger than what you began with?

  ✔ *Taking away*: Is your answer smaller than what you began with?

✔ *Real-life problems*: What would you have guessed, roughly?

✔ *Probability*: Is your answer between 0 and 1?

You can probably – and should probably – come up with a dozen similar questions about whether an answer makes sense. Before you even begin work on a question, you can start thinking about what criteria your answer has to satisfy.

Eliminating Wrong Answers

If you have a multiple-choice question – I don't pose many of those in this book, but some numeracy tests use them extensively – you can use the 'making sense' test on each of the answers you're given. Quite often, one (or more) answer is quite obviously wrong. You can cross these out without fear – and then decide between the answers that remain.

Explaining to yourself *why* you think an answer is wrong is a good idea. Saying 'That's too big' or 'That ends in the wrong number' can help train you to spot wrong answers when you generate them! (No matter how smart you are, you *will* come up with wrong answers. You're allowed to get things wrong; it's part of the learning process.)

Working Backwards

Once you have an answer, you can often check whether it's right by working back from it to get the question.

For example, if your question asked you when a train would arrive after a given travel time, you could use your answer and the start time from the question to work out the travel time – and check you get the same number as the question. Alternatively, you could use your answer and the travel time to figure out when the train left – and again check you get the same thing.

If you get the same answer as was in the question, you can probably pat yourself on the back and believe in your answer; if you get a different answer, running through it one more time to see if you can find a mistake is probably worthwhile.

Re-reading the Question

When you have an answer, checking that what you've written down answers the question is *always* a good idea. For example, if a question asks for an answer in minutes and you give it in hours, you'll lose marks, even if you've done the hard part correctly. This punishment may be harsh, but that's the way things go.

You're not a politician. Answer the question the examiner asks you!

Using the Information

With the exception of some questions involving graphs and tables, every piece of information the question gives to you is usually there to be used on the way to your answer.

If you're stuck on a question and don't know where to begin, you can start by writing down all the information – neatly! Give everything a name and then think about how you can combine it.

If you get to an answer and find there's some information you haven't used, you can also have a think about how you might have done so.

Rounding Roughly

In Chapter 4, I show you how to come up with an approximate answer to a complicated sum. You may be asked to do just that in an exam question, but it's also a useful tool if you want to know roughly what your answer should look like.

By coming up with a rough answer, you can tell whether the sum you've worked out the hard way is on the right lines. For example, if you get an answer of 0.253 the hard way and your rough answer is 7,000, you can be pretty sure something has gone wrong. If your rough answer was 0.2 or 0.3, you'd be a lot happier with it!

Normally, if your rough answer is more than half of your accurate answer and less than double it, you can take that as a good sign. It doesn't mean you're definitely right, but you're at least in the right ballpark.

Trying Another Way

One of my favourite analogies is that solving a maths problem is like driving to a destination: sometimes you don't know where you're going, but there's always more than one way to get there.

Which means, getting lost doesn't indicate you're a bad driver! It just means it's time to find a different route to where you want to go. The nice thing about maths is that you can always go straight back to the beginning – or, if you've laid your work out neatly, any point on the way.

Getting stuck doesn't mean you're useless, just that the thing you tried didn't work. Try something else. Make mistakes fearlessly – that's how you learn.

Using Your Common Sense

Guesswork isn't really a good way of making sure you get the right answer, but using what you know about the real world to check your answers are sensible is . . . well, sensible!

For example, if you're asked to work out the weight of a car and your answer is a few grams (rather than a tonne or so), you know you've gone wrong. If you're looking at the height of a door and you get an answer of several kilometres, that's really not a plausible answer. If you're working out the cost of a pint of milk and it comes to £50, it had better be unicorn milk!

Before you answer a question that's based in real-life, have a think about what you'd expect the answer to be – just roughly. Then you can decide whether what you've worked out is a realistic answer.

Taking Your Time

Whatever you do – don't rush unless you really have to!

Doing maths is much harder when you're stressed. When you sit down to study, it's normally much better to do one question at your own pace rather than trying to cram five questions into a specific time slot. Especially if you're not that confident, trying to speed through questions before you understand what's going on is a sure-fire way to make yourself miserable and dejected.

Instead, take as long as you need and get as much help as you need. The aim of studying is to learn the subject not to achieve some crazy goal of answering every question in the book within 20 minutes.

No global backlog of maths questions exists and you don't need to answer as many as you can as quickly as you can to save the planet! Slow and steady wins the race.

There's one main exception to this advice: if you're studying for an exam and you know time is going to be tight – *then* you can do some speed training. But not until you know how to do the sums first – and that means doing them slowly to begin with.

Drawing a (Big!) Picture

I go through pitched battles with my students trying to get them to draw pictures of anything that could be even slightly relevant to the question they're trying to answer.

For a good picture you need:

- ✔ *A title, if relevant.* Two days from now, you won't know what the point of the picture was. Make life easy for future-you!

- ✔ *A pencil.* Pen drawings look nicer, but they're harder to correct. Do everything in pencil first and then go over it in pen when you're happy.

> ✔ *Clear labels*. Everything that can be labelled, should be.
>
> ✔ *Plenty of space*. A cluttered diagram is hard to read.
>
> ✔ *Colours*. If your graph or whatever is at all complicated, using several colours will help you keep things straight in your mind – and help you remember what's what.

If you find your diagram is getting messy, or if you make a mistake that's too big to fix, don't be afraid to scrunch it up – or better, cross it through neatly – and start another one on a fresh page.

Checking Your Units

In 1999, one of NASA's Mars Orbiters disintegrated on its way through the Martian atmosphere. Hundreds of millions of dollars and years of work burnt up in a matter of seconds.

They eventually figured out what had caused the accident: one of the programs controlling the spacecraft used imperial units such as inches; another used proper, grown-up units like metres. The two programs had a . . . bit of a communication problem, leading to catastrophe.

Whoops.

Now, the good news is, even if you get every question in the exam wrong, you're not going to have to justify the failure of a very expensive mission to some very angry budget people. You can even take some consolation from the fact that rocket scientists are just as capable of getting things monumentally wrong as everyone else.

But, just to be on the safe side, make sure you're using the right units every time you work through a question.

Chapter 20

Ten Tips for Remembering Your Number Facts

*P*ersonally, I don't like the way that numeracy tests put so much emphasis on quick-fire number skills rather than the much more useful things like solving problems and using calculators and computers. Don't get me wrong: mental arithmetic is handy and anyone with maths skills should be proud of them – I just don't think they're the be-all and end-all.

All the same, we're stuck with the tests we've got, and it's worth learning your number skills as quickly and efficiently as possible. The quicker you have complete control over your number facts, the less time you'll have to spend learning them. More importantly, knowing your number facts also means you'll find the other sums come much more easily and quickly.

In this chapter, I give you ten things you can do to make learning your facts a breeze.

Playing Games

Games are a really useful way to learn anything at all. They take the chore element out of learning and turn it into something a bit more fun. Playing a number facts game using cards or the computer feels like less hard work than writing long lists of sums. I strongly recommend using games as a learning tool – let me know if you find any good ones!

In the nearby sidebar I provide a list of online games. I particularly like the BBC Skillswise site because it's easy to navigate and – according to my students – fun to use. It's not the only one, though: all of the sites listed in the 'Playing Games Online' sidebar can be useful – try a few and see which one is best for you.

Playing games online

Literally hundreds of websites offer games to help you learn maths facts and techniques. Some are fun, some are more serious and others fall in-between. Here are some of my favourites:

✔ www.mangahigh.com – Manga High is a well-designed maths games and lessons website. It covers a wide range of topics and has probably the most exciting maths games on offer.

✔ www.bbc.co.uk/skills wise/numbers – The BBC site is a bit more focused on adult numeracy than Manga High, and organised with fact sheets and games to help you keep on top of your topics.

✔ www.mymaths.co.uk – MyMaths is a subscription site that concentrates mainly on GCSE and A-level maths. That said, it does have some free sample games you can play to practise some of your skills.

✔ www.mathmotorway.com – Math Motorway is simple but brilliant: answer questions on adding, taking away, multiplying and dividing as fast as you can. Whenever you get one right, your car moves forward. Can you win the race?

✔ www.sumdog.com – Sumdog is a great free maths games site with all manner of games, from penalty shoot-outs to *X Factor*-style audition games. The questions respond to how well you're doing with each topic; once you've mastered something, you won't see any more questions related to that subject. A really good site!

Flashing Cards

For some reason, flash cards haven't really taken off in the UK. In America, using cards is *the* way to study. You write your key facts down on postcard-sized bits of cardboard (question on one side, answer on the other) and quiz yourself relentlessly.

I recommend reading the question aloud, answering it and then checking whether you're right. If your answer is correct, put that card to one side; if you're not happy with your answer, put the question to the back of the pack. Once you've answered all of the questions correctly, give yourself a pat on the back (or a more substantial treat if you prefer).

Working your way through a pack of flash cards can be a bit tedious, but this method is really effective. You'll be surprised by how quickly you can go from knowing a few of the answers to knowing them all.

The more your practise anything, the better, and more confident, you get. Flash cards are a really effective way of practising the things you need to practise.

Sticking Stickies

Whenever I was studying for exams, the walls of my room would be *covered* in notes and pictures so that I couldn't avoid seeing what I was meant to know.

One of my students came up with the idea of putting sticky notes on every door in her house with a question on one side and the answer on the other – before she could go through a door she had to answer a question. (She got a fantastic grade, so clearly her method worked!)

You may not feel like decorating your entire house with maths notes, which I understand. What you could do instead is put a few yellow sticky notes with things you want to remember on surfaces you look at frequently – your computer or your bathroom mirror, perhaps.

When you've learnt everything on a sticky, discard it and make a new one with the next thing you need to learn.

Counting on Your Fingers

Some maths teachers are very strongly against using your fingers to work things out, reasoning that it's like reading words letter by letter – it slows you down unnecessarily.

I'm not quite so dogmatic – anything that gets you the right answer is fine by me – but I'd much rather you learn your times tables and adding tables by heart. It looks much better if you answer questions without needing to count!

So, if you *are* going to use your fingers to work things out, I'd like you to do one thing for me: whenever you do so, say the sum out loud (for example, 'nine add six is fifteen'). Repeating the sum out loud reminds your brain that this is something to remember.

Tricking out the Nines

Here's a nice, easy way to work out nine times anything (up to ten). Here goes:

1. **Take one away from the number you're timesing by and write it down.**

2. **Take your answer to Step 1 away from 9, and write this number down to the right of your answer from Step 1.** That's it!

So, to do 9 × 7, you'd say, 'I'm timesing by 7, so I write down 6; 9 – 6 = 3, so I write that next to it . . . the answer is 63.'

You can check to see if your answer to a nine times table sum is plausible by adding up the digits of your answer. If you get a number that's not in the nine times table, you've made a mistake.

Unfortunately, this method only works for the nine times table.

Tricking Out Other Big Numbers

Here are some tricks for your other 'big' times tables: 5, 6, 7 and 8.

Tricks of six

To multiply by six, follow these steps:

1. **Double your number.**

2. **Multiply this answer by three.**

That's it! It doesn't even matter which way around you do the steps. So, to work out 12 × 6, you do 12 × 2 = 24, then 24 × 3 = 72.

Straight to eight

Here's a similar trick for multiplying by eight:

1. **Double your number.**

2. **Double it again.**

3. **Double your answer one last time.**

There! So, to figure out 15 × 8, you can do 15 × 2 = 30, then 30 × 2 = 60 and 60 × 2 = 120.

What about seven?

Seven is harder, I'm afraid. However, if you can multiply by six (using the trick described above), you can multiply by seven like this:

1. **Multiply your number by six.**

2. **Add your number on to your answer.**

So, to work out 7 × 13, you'd do 6 × 13 = 78, and add on the 13 to make 91.

This times table recipe is the only one in this section in which the order matters – you have to do the × 6 part first.

Five alive

For my last (multiplying) trick, I show you how to multiply by five. Follow these simple steps:

1. **Multiply your number by 10.**

2. **Halve the answer.**

So, to work out 15 × 5, you figure out 15 × 10 = 150, then halve your answer to get 75.

The order in which you do these two sums is irrelevant. If you prefer to halve the number first and then times by ten, that process works just as well.

Breaking Down and Building Up

Dividing tricks are just like multiplying tricks. This may come as no surprise because multiplying and dividing are opposites of each other.

Figuring out eight

Here's a very easy way to divide by eight:

1. **Halve your number (divide it by two).**

2. **Halve it again.**

3. **Halve it one last time.**

So, to work out 360 ÷ 8, you halve it (360 ÷ 2 = 180), then again (180 ÷ 2 = 90) and one last time (90 ÷ 2 = 45).

Surprised by six

If you notice that the recipe for dividing by eight simply involves reversing the steps you follow in multiplying by eight, you'll be delighted to hear that the same is true for dividing by six. You just halve the number and divide the answer by three.

So, to work out 114 ÷ 6, you halve 114 to get 57, and divide that by 3 to get 19.

You can do these steps in either order.

Nailing nine

Dividing by nine is no more taxing than the others: you just divide by three and then three again.

To work out 117 ÷ 9, you divide by 3 to get 39, and divide 39 ÷ 3 = 13.

Finally fives

The recipe for dividing by five is just the opposite of multiplying by five: you double your number and divide by 10.

If you want to know 235 ÷ 5, you double 235 to get 470; then you divide by 10 to get 47.

You *can* do these steps either way round, but only if you're happy working with decimals! If you divide 235 ÷ 10, you get 23.5. If you look at that and think 'I could double that!', then do so; but if you're not sure, you can avoid the decimal by multiplying first.

Working from What You Know

You can split up any multiplying sum into smaller multiplying sums and add the answers together! For instance, to work out 12 × 12, you could split up the sum in several ways:

- ✔ If you know 12 = 2 × 6, you can work out 12 × 2 = 24, then 24 × 6 = 144.

- ✔ If you spot 12 = 10 + 2, you can work out 10 × 12 = 120 and 2 × 12 = 24 and add the answers together: 120 + 24 = 144.

Powering through the Slumps

When you're learning *anything*, you have some days when you feel hopeless and that you'll never get it. Horrible and disheartening as these days are, you can get through them – promise!

You can power through the slumps in lots of ways. My favourite is to look back at the things I was learning two months ago and think, 'Good grief, did I really struggle with that stuff? It's so easy now.' Sometimes you're so focused on how far you still have to go that you forget how far you've come!

You can also try changing your scenery (I decamp to the coffee shop if I'm struggling to write; if nothing else, a cappuccino will cheer me up!) or doing some exercises. Just doing something slightly different for a little while can pay big dividends.

Learning from Your Mistakes

Always learn from your mistakes. Most people look at mistakes as nuisances and things to avoid – and, obviously, you shouldn't go out of your way to make them! However, I make mistakes, you make mistakes, Brian Cox makes mistakes (have you heard D:Ream's first album?); the trick is to learn from them. That's why Brian Cox is now a physicist, not a pop star.

Whenever you notice you've made a mistake, make a note of what you should have done. (You could even put it on a flash-card or a sticky to review later.)

Being upbeat about mistakes is the key thing; just accept that they happen. They don't mean you're hopeless; they're just bumps in the learning curve that you'll get over with a little work.

Chapter 21

Ten Top Tips for Getting Things Right

1 think most people have traumatic memories involving getting stuff wrong in school – from saying something in class that you didn't realise was idiotic to wearing the wrong jacket.

One of the good things about being a grown-up is that nobody's going to mock you for making mistakes. In fact, my attitude is that if you're not making mistakes, you're probably not stretching yourself! So I give you a quiet round of applause every time you mess up.

All the same, you don't go *looking* to make mistakes, especially in tests. They're part of the learning process, but obviously you want to minimise them in the end. This chapter gives you ten ways to help make sure you come out of the final exam smiling even if you are wearing the wrong jacket!

Keeping in Practice

If you do something over and over again, you get better at it. That's why sports teams train, why musicians practise and why you should make sure you do at least a little bit of maths every day.

When Jerry Seinfeld was a young comic without much material, he made the decision to write something every day, no matter how little. He got hold of a cheap calendar and hung it somewhere he couldn't miss it. Every day he wrote, and drew a cross over the day. After a while he noticed that the crosses were forming a chain – and it'd be a shame to break the chain, wouldn't it?

And the upshot was Seinfeld went from being an obscure up-and-coming comedian to one of the most successful comedy writers and performers of his era.

 Get a cheap calendar. Hang it somewhere prominent. Put a cross on it every day you do some maths. Don't break the chain!

Working on the Easy Stuff

I can't think of many things more frustrating than sitting in an exam, looking at a question and thinking, 'I used to know this, but I can't quite remember it.'

The only way to avoid it being in this situation – and there's no cast-iron guarantee, even here – is to make sure you practise the things you already know. I recommend running through some familiar topics at the start of a study session, like a warm-up, to give you a chance to ease in gently and give you a nice confidence boost as well!

Working on the Hard Stuff

Sometimes, when I'm working with a student, the lightbulb suddenly goes on. Their eyes open wide, they say 'Oh! This is easy!' and we both feel fantastic.

That eureka moment can happen at any time, often when you're not directly working on maths – I've had flashes of inspiration come to me in my sleep, in the shower and even in an exam (that one was just a few hours too late for me to do anything about it, though). It's as if some invisible elves show up, whisper a clear explanation and it all makes sense.

The thing is, inspiration like this only comes when you've been working on something difficult and haven't come up with an answer. If you make a habit of spending a few minutes of each study session looking at something you don't quite get, you make it much more likely that the inspiration elves will pay you a visit.

Laying It Out Neatly

Every so often, I see a student's work that resembles an accident with alphabetti spaghetti. If they've reached the right answer, it's not necessarily a problem . . . but it's impossible to put right if something goes wrong.

If you keep your ideas straight, finding – and correcting – mistakes is so much easier. Here are a few things you can do:

- ✔ If you're using lined or squared paper, use the lines and the squares to help you. Put one number (at most!) in each space.

- ✔ Give yourself plenty of space. If you're asking yourself, 'Can I fit the next question in there?', the answer is no.

- ✔ Cross things out neatly. Scribbles are cathartic, but make understanding your notes later really difficult.

Keeping Your Ideas Straight

Nothing's worse than seeing a page full of numbers and not knowing what any of them mean. (Well, there may be worse things. But it's still pretty bad.)

Remember – you're not just writing things down to get the right answer. You're writing things down so that you can revise from your notes later and see what you did. A mess of meaningless numbers is no use to anyone, even if you get the right answer.

I recommend copying down the question before you start working on it, and then whenever you write a number down, write what it is. For example, don't just write '2:15'; write

'Start time: 2:15'. That's why I always make a point of labelling the Table of Joy – it tells you what's what when you come to look at it again!

Making a Plan

When you're planning a route from one place to another breaking it down into smaller, logical steps on the way is a good idea. So, too, can you navigate through a maths problem by making a plan of action.

Many maths questions – especially the Table of Joy questions that make up a big proportion of any numeracy test (see Chapter 7 for the lowdown on this marvellous grid) – follow a similar structure, which makes it easy to say, 'I'll do this, then that, then the other thing'. That's why it's possible for me to write recipes for them! If you can look out for these structures and notice how they work, you can avoid at least some blind alleys and make sure you end up at the right answer.

Treating Yourself Well

If you want to train a puppy to behave itself, you give it treats and attention when it does the kind of thing you want it to. Over time, it associates doing what it's told with pleasurable experiences and eventually Fido will sit or beg or play dead or herd sheep whenever you give the word.

Now, I'm not calling you a puppy – that would be mean – but the human brain and the dog brain aren't really all that different in one respect: the human brain responds in much the same way to rewards. You can use this to your advantage!

For example, whenever I finish writing a chapter, I treat myself to a good cup of coffee and a slice of carrot cake. Some mornings I wake up and think, 'I don't want to write today', but that's quickly overtaken by, 'If I get a chapter done, I can have some cake!' – so I start writing.

You can do the same thing with maths – give yourself a small treat whenever you do something you're proud of.

Good puppy!

Sitting Up Straight

I'm writing this hunched over my laptop on a crowded train, so I'm probably not in the best position to tell you that posture is important for top performance. It sounds a bit like new-age nonsense, but sitting up straight can make a big difference to how you fare in your studies.

Your brain speaks fluent body language. If you sit upright, with your shoulders back and head up, your body says, 'I'm confident and capable of dealing with anything life throws at me', and your brain says, 'Bring it on!' If you crunch up and cower over your desk, your body says, 'Don't bother me, I don't want to interact', and your brain says, 'Sorry, closed for business'.

If you start to feel a bit overwhelmed with your studies, take a few seconds to notice how you're sitting – try to adjust yourself slightly and adopt a more commanding posture.

Breathing Like a Rock Star

I used to suffer from panic attacks – they're horrible and I wouldn't wish them on anyone. One of the tricks I was taught for dealing with panic attacks involved diaphragmatic breathing.

The diaphragm is a muscle somewhere in your lower chest. Instead of taking shallow breaths into the top of your lungs as you do with normal breathing, with diaphragmatic breathing you take the breath as far down into your lungs as possible and then breathe out slowly. Singers use diaphragmatic breathing before they go on stage: it has the twin benefits of helping lung capacity so they can sing better, and calming them down so stage-fright doesn't hit so hard.

Here's how to do it:

1. **Put one hand on the top of your chest and the other hand on your belly.**

2. **Breathe in as deeply as you can, trying not to move your upper hand.** Your lower hand should move out as your lungs fill with air.

3. **Breathe in for a count of seven, but don't hold your breath.**

4. **Breathe out very slowly for a count of 11.** Don't worry if you can't manage all the way up to 11 – just breathe out for as long as you can and make a note to breathe more slowly next time.

5. **Follow this breathing pattern for a minute or so and you'll feel your heart rate drop and your head start to clear.** A clear head makes maths a lot easier.

I can't emphasise enough how helpful this type of breathing is to me when I feel panicky or stressed or even just out of sorts. Taking a few minutes to breathe deeply and get some oxygen into my body has saved more than one day from being a complete write-off for me. Coupled with positive self-talk, diaphragmatic breathing is a tremendously powerful weapon against the panic-monsters, either when you're revising or just before the test.

Talking Yourself Up

The way you talk to yourself makes an incredible difference to how you perform. If I tell myself 'I can't dance and it's no fun anyway', I'm likely to have a miserable and embarrassing time trying to learn to salsa. If I tell myself, 'Well, it's exercise, and nobody's watching me, and making mistakes is perfectly normal', then suddenly I'm having more fun than I have any right to.

You can do the same with maths. Instead of saying, 'Maths is stupid and difficult. I can't do it; I'm not a maths person and my brain doesn't work that way', or anything else that's demoralising or nonsense, try saying, 'I'm smart, and I can figure it out. I have an amazing brain – I just need to train it a bit more.' Best of all, try saying: 'Ah, I got that wrong. That's interesting. I wonder why?'

Index

FOR DUMMIES®

Making Everything Easier!™

UK editions

FOR DUMMIES

Making Everything Easier! ™

UK editions

BUSINESS

Bookkeeping FOR DUMMIES

978-0-470-97626-5

Persuasion & Influence FOR DUMMIES

978-0-470-74737-7

Starting & Running a Business ALL-IN-ONE FOR DUMMIES

978-1-119-97527-4

REFERENCE

British Politics FOR DUMMIES

978-0-470-68637-9

DIY FOR DUMMIES

978-0-470-97450-6

Dad's Guide to Pregnancy FOR DUMMIES

978-1-119-97660-8

HOBBIES

Growing Your Own Fruit & Veg FOR DUMMIES

978-0-470-69960-7

Keeping Chickens FOR DUMMIES

978-1-119-99417-6

Beekeeping FOR DUMMIES

978-1-119-97250-1

Asperger's Syndrome For Dummi
978-0-470-66087-4

Basic Maths For Dummies
978-1-119-97452-9

Body Language For Dummies,
2nd Edition
978-1-119-95351-7

Boosting Self-Esteem For Dummi
978-0-470-74193-1

British Sign Language For Dumm
978-0-470-69477-0

Cricket For Dummies
978-0-470-03454-5

Diabetes For Dummies, 3rd Editio
978-0-470-97711-8

Electronics For Dummies
978-0-470-68178-7

English Grammar For Dummies
978-0-470-05752-0

Flirting For Dummies
978-0-470-74259-4

IBS For Dummies
978-0-470-51737-6

Improving Your Relationship
For Dummies
978-0-470-68472-6

ITIL For Dummies
978-1-119-95013-4

Management For Dummies,
2nd Edition
978-0-470-97769-9

Neuro-linguistic Programming
For Dummies, 2nd Edition
978-0-470-66543-5

Nutrition For Dummies, 2nd Editi
978-0-470-97276-2

Organic Gardening For Dummies
978-1-119-97706-3

**Available wherever books are sold. For more information or to order direct go to
www.wiley.com or call +44 (0) 1243 843291**

Notes

Notes